Sonic Persuasion

STUDIES IN SENSORY HISTORY

SERIES EDITOR
Mark M. Smith, University of South Carolina

SERIES EDITORIAL BOARD
Martin A. Berger, University of California at Santa Cruz
Constance Classen, Concordia University
William A. Cohen, University of Maryland
Gabriella M. Petrick, New York University
Richard Cullen Rath, University of Hawaiʻi at Mānoa

Sonic Persuasion

Reading Sound in the Recorded Age

GREG GOODALE

UNIVERSITY OF ILLINOIS PRESS
Urbana, Chicago, and Springfield

© 2011 by the Board of Trustees
of the University of Illinois
All rights reserved
Manufactured in the United States of America
1 2 3 4 5 C P 6 5 4 3 2
♾ This book is printed on acid-free paper.

Library of Congress Cataloging-in-Publication Data
Goodale, Greg, 1966–
Sonic persuasion: reading sound in the recorded age /
Greg Goodale.
p.   cm. — (Studies in sensory history)
Includes bibliographical references and index.
ISBN-13: 978-0-252-03604-0 (hardcover: alk. paper)
ISBN-10: 0-252-03604-2 (hardcover: alk. paper)
ISBN-13: 978-0-252-07795-1 (pbk.: alk. paper)
ISBN-10: 0-252-07795-4 (pbk.: alk. paper)
1. Sound—Recording and reproducing—United States—History.
2. Sound recordings—Social aspects—United States—History.
3. Persuasion (Psychology)
I. Title.
TK7881.4.G643   2011
302.2—dc22   2010041880

Contents

List of Illustrations   vii
Preface   ix
Acknowledgments   xiii

1. Reading Sound   1
2. Fitting Sounds   16
3. Machine Mouth   47
4. The Race of Sound   76
5. Sounds of War   106
6. On Sound Criticism   132

Notes   155
Index   183

Illustrations

2.1: Joseph Keppler Sr., "The Bosses of the Senate" (1889)   27
2.2: Bernhard Gillam, "Patching Up the Old Ex-Champion" (1892)   31
2.3: McKee Barclay, "Pianissimo Teddy!" (1910)   38
2.4: J. L. De Mar, "Roosevelt as a Schoolmaster" (1906)   40
2.5: William A. Rogers, "Congress on His Hands" (1903)   41
2.6: Samuel D. Erhart, "Baby Kiss Papa Good-bye" (1909)   43
3.1: "Molloch" from *Metropolis* (1927)   57
3.2: Umberto Boccioni, *The Noise of the Street Enters the House* (1911)   59
3.3: Sybil Andrews, *Sledgehammers* (1933)   66
4.1: Benny Goodman's studio (1934)   89
4.2: Chick Webb, Artie Shaw, and Duke Ellington (1937)   90

Preface

The genesis of this project lies in Lawrence W. Levine's single-word response to a question I posed about obscure historical sources. To complete a graduate class in cultural history, I attempted to make sense of thousands of letters written to console Ida McKinley after the assassination of her husband in 1901. As a nascent scholar, I was trying without success to fit these letters into simple categories. When I asked Levine what to do with the letters, he cryptically answered, "Listen." Of course, the correspondents were dead, and in 1901, few Americans had access to sound-recording devices. Surely, Levine didn't mean that I should literally listen to scraps of paper. What Levine meant was to listen metaphorically to the sources. He taught his graduate students that as scholars, we are not trained to listen—we are trained to read. After five centuries of the book, scholars have become accustomed to perceiving the world only through the lens of reading. Even when we study speeches or the lyrics to popular songs, we rarely study the sounds of voices and music. Rather, we convert sounds into words on a page.

The legacy has left scholars in the humanities with a host of visual metaphors for thinking. Things "appear" to be, subjects "see" that, history is "unveiled" to us. It is a rare metaphor that compares the other senses to the acquisition of knowledge. And yet, we learn from taste, touch, smell, and sound as well as from sight. Inattention to all five senses, a problem that many scholars are beginning to address, leaves our understanding of both history and the present disabled and leaves us prey to the manipulations of those who understand the persuasive powers of the nonthought senses. Sound is one sense that carries great rhetorical force in and of itself. Yet, scholars have inconsistently thought about it. Perhaps this is why Levine asked his students to listen.[1]

Levine's sound advice to me was not the first time he looked to the sonic for guidance. In class, he was fond of anecdotes like the one about immigrants to America who went to the talkies for the first time and knew by the soundtrack that the bad guys (usually Indians) would soon appear on the screen. The post–Civil War blues was a favorite topic. A recording of Levine analyzing an old blues tune has him saying, "You could learn a lot just from *listening carefully, reading carefully.*"[2] His comparison of listening to reading was and remains provocative. Can we read through our ears? Yes, Levine would have argued, and so I, too, take on the visual metaphor of *reading* sound when in fact I am describing the experience of listening closely. He also left a tantalizing suggestion about the sound of oratory in an essay about Shakespeare in American popular culture: "The generations of people accustomed to hearing and reciting things out loud—the generations for whom oral recitation of the King James version of the Bible could well have formed a bridge to the English of Shakespeare—were being depleted as America entered a new century." To this, Levine attached an endnote: "The relationships between recitation of the King James Bible and performances of Shakespeare and between the transformation of nineteenth-century religious style and the transformation of Shakespeare need further thought and research."[3] Though Levine was relying on the written record, he had listened to the sounds of American speech correctly: a transformation from what we would perceive as florid recitations of the Bible or Shakespeare to a more simple oratorical style did indeed occur at the turn of the twentieth century.

Levine's sonic examples pervade this project: criticism by listening rather than seeing, the search for sonic evidence, and the willingness to eavesdrop on historical shifts. Sadly, however, few scholars have made prolonged forays into sonic criticism. Unlike the sustained and multifaceted turn toward visual studies in the 1970s, scholarship about sound has just begun to blossom. The vocabulary of its practitioners remains a mishmash of theoretical perspectives and its guiding principles a polyglot of intellectual traditions. To be sure, a few heavyweights have waded into the field. Footnote watchers will find Theodor W. Adorno, Roland Barthes, Pierre Bourdieu, Jacques Derrida, Marshall McLuhan, and Walter J. Ong cited in this study. Yet, there are no coherent approaches and no sonic methodologies that bridge the divides among disciplines in the humanities. Words like *icon* and *gaze* help constitute a lingua franca among visual scholars, a vocabulary for whom sonic studies has no parallel. Levine could have used such a lexicon. His last book, *The People and the President: America's Conversation with FDR,* written with his wonderful wife, Cornelia, employed scholarship from the 1930s about the speed of speech to explain the rhetorical force of Franklin Delano Roosevelt's fireside chats. It wasn't just words that FDR used to change America; it was also cadence that made him such an effective communicator. Of course, twenty-first-century scholars who read transcripts of FDR's chats won't find in those *words on a page* the pace at which the president spoke the words. The Levines relied on their ears and on scholarship from the 1930s to *listen* to FDR and in doing so discovered that he spoke incredibly slowly. Every word of his "day of infamy" speech, spoken at the snail's pace of eighty-eight words per minute, fell upon his listeners like a sledgehammer. But *cadence*, from the Latin "to fall," is a word rarely seen in criticism of speech anymore.[4] By listening closely, we will remind ourselves of the importance of sonic concepts like cadence and discover histories that have so far remained invisible.

Acknowledgments

I would like to thank three groups for their assistance in helping to research, write, and publish this book. First, I thank the librarians and archivists at the University of Illinois libraries, Snell Library at Northeastern University, the Boston Public Library, Mugar Library at Boston University, the Boston Athenaeum, and the Gilmore Music Library at Yale University. Librarians and archivists at these institutions dug up the most extraordinary materials for me to listen to and read.

Second, I thank the many scholars who have read and heard chapters of the book over the past few years. These include Mark Smith, Joshua Gunn, Mirko Hall, Thomas Rickert, Geoff Carter, Ryan Blum, Dan Larson, Eric Jenkins, Ian Reyes, Cara Finnegan, David Cisneros, Charles Morris III, Cara Buckley, John Louis Lucaites, Eric King Watts, Jeremy Engels, Dana Hall, Mattea Garcia, Erik Johnson, Brett Ommen, Karlyn Kohrs Campbell, and Adrienne Hendee Jacobson. The best part of writing is the scholars one gets to engage with and I have derived much enjoyment from working on this book.

Last, I thank the staff at the University of Illinois Press who have done such a professional job of getting the book from concept to publication. These include Kendra Boileau, Danny Nasset, Jennifer Clark, Joseph Peoples, Mary Lou Kowalski, Jennifer Reichlin, David Drummond, Mary Wolfe, and Roberta Sparenberg. I would recommend to any author that they find a way to work with this group of professionals.

Sonic Persuasion

1

Reading Sound

Undergraduate students often encounter Franklin Delano Roosevelt's first inaugural address during their academic adventures. Political scientists, American historians, and rhetorical critics, in particular, like to use this speech. The oration is certainly historic, not to mention interesting and easy to read. The address also features a line, "the only thing we have to fear is fear itself," that has been woven into the fabric of public memory. Unfortunately, the line is no longer as powerful as it once was. Inevitably, when the speech and that sentence are discussed in undergraduate classes, they are discussed as *words on a page*. Indeed, the address is a text composed of powerful words ranging from appeals to fear, to military and Christian metaphors, to attacks against moneychangers.[1] But to say that only words constitute the text is to miss part of the reason why that speech resonated so deeply with Americans in 1933. FDR did not present that famous sentence in a monotone voice, nor did he vocalize the words like a thespian we might berate today as overacting. Instead, he raised his voice between "the only thing we have to fear" and the word "is." Then he paused for an excruciatingly long moment before finally finishing the sentence. The pause marked Roosevelt's close attention to the sound of his voice, his understanding of the power of radio, and his awareness of what the pause would do to his audience.[2] As a result, FDR captured the hearts of a nation.

The educator John Erskine understood what FDR was doing. In a text about using the radio voice to persuade, Erskine notes, "We need attend to only two points: first, how to persuade our audience to come in; second, how to prevent it from walking out. There are many reasons why it may come in, but it will stay only because the performance seems worth while, or because we have locked the door on it."[3] FDR's pause captured his audience's attention. In contrast to those who have forgotten that FDR's words had sound, radio-savvy communication professionals understood the power of a silent pause in the 1930s. Arch Oboler wrote for, produced, and announced the popular horror program *Lights Out*. In 1939, Oboler explained to *New York Times* readers that "often a silence or a pause between words is more important than the spoken word, because the listener, in the mind's eye, during the pause, is contributing to the play. His imagination gets a chance to work; he is experiencing the play more emotionally."[4] By ending the first phrase with a question about what we have to fear, Roosevelt encouraged his listeners to imagine and to experience his words emotionally, fearfully. "What do we have to fear?" countless Americans must have thought during that long pause in 1933. Then, relieving his audience of the full tension of fright, FDR answered only fear itself. Did Roosevelt know what he was doing? He must have. In 1932, the *New York Times*' radio columnist, Orrin E. Dunlap Jr., described the potential power of the political vocalist on the radio: "There are no torches, gestures, bands in regalia, bunting or mob enthusiasm and emotion to supplement the oratory and help to hold attention."[5] Holding the attention of millions of listeners required purely vocal devices—in FDR's case a rising voice, a pause, and an answer. The lived moment, the moment that the immediate audience heard, was and remains a far more powerful moment than the clever use of repetition that twenty-first-century students find in textbook histories. The written word cannot convey the power of the spoken word on that cold March day at the beginning of Roosevelt's extraordinary presidency, but the sonic record can.

To overstate the impact of this sonic manipulation would be difficult. In an era marked by the proliferation of instruments of sound without sight—the telephone, the phonograph, and the radio—the power of the voice grew to ever-larger proportions as an agent for changing public opinion. Roosevelt understood the power of invisible speech and employed it forcefully when he paused, emphasized, or intoned. Roosevelt's cadence, for example, was extraordinarily slow, particularly during his most important speeches and fireside chats.[6] Adolf Hitler, to provide another example, did not just say angry words. He enveloped his listeners in a national voice and in doing so helped to drive the world into a six-year conflagration. Moreover, it is no accident that the greatest media hoax in American history occurred over the radio in 1938. Orson Welles's broadcast of *War of the Worlds* during the Columbia Broadcasting System's *Mercury Theatre* was intended to manipulate. Listeners were frightened by sound effects that producers and actors employed to portray an alien attack on the United States. Actors sounded like panicked reporters or government officials (one actor adapted the cadence of FDR's fireside chats), while Welles's staff put on all the sonic tricks they could muster to imitate sonic expectations about an alien attack. The audience's panic is enlightening. Twenty-first-century listeners are more likely to laugh at the campy sounds. The difference illustrates just one of the many arguments about sound that I make in this book: sonic expectations derived from the "period ear" change over time.[7] Yet, it is only in the past decade that scholars have begun to examine the power that sound—sonic persuasion—has over audiences who have not been trained to understand that a pause, or an intonation, or even a noise can make a forceful argument.

Sounds Invisible

There is an ancient history of emphasizing vision rather than hearing in the Western tradition. Plato, philosopher and opponent of rhetoric, posited a visual metaphor twenty-four hundred years ago that is still taught in academia. It is the man in a dark cave (a metaphor for ignorance), who only perceives shadows cast on the walls by a small fire. Once exposed to the sunlight (a metaphor for knowledge), his frame of reference is inalterably changed, and he becomes *Homo sapiens,* thinking man. Plato disdained sound because unlike light, noise distracts from knowledge absent the spoken word. Likewise, because it is emotionally appealing, music prevented lovers of knowledge from knowing the world.[8] To ignore affect, however, is to ignore rhetorical force. Parallel to his condemnation of rhetoric, Plato also rejected voice in favor of logos and in doing so lost sight of the persuasive impact of tone, silence, and volume.[9] In short, Plato was blinded by the sunlight. He was thus prevented from seeing that sound conveys much meaning.

Though Enlightenment philosophers and scientists like Francis Bacon and Robert Hooke tried to make sound the equal of sight in their observations of natural phenomena, the sonic remained secondary. Thus, science, though it should be founded on all of the senses, has always been closely wedded to visual examination.[10] The dominance of vision proposed by Plato and maintained by scientists continues to blind scholars to new avenues of understanding history, culture, and other domains of study in the humanities. Even in psychoanalysis, which should dig beneath the visual, practitioners have focused on the gaze rather than the voice.[11] Though critics of words and images have produced outstanding analyses of history and culture, Plato's blind spot limits scholarly thinking. As the media theorist Marshall McLuhan notes, this blindness runs deep: "Western man thinks with only one part of his brain and starves the rest of it. By neglecting ear culture, which is too diffuse for the categorical hierarchies of the left side of the brain, he has locked himself into a position where only linear conceptualization is possible."[12] Entirely new ways of thinking, McLuhan proposes, are possible if only scholars address the neglect of ear culture.

Decades after McLuhan's call, our understanding of voices, music, and noises remains disabled, even as scholars in the humanities valiantly attempt to cure our blind spot. Historians of phonographs and radios often lament the absence of aural understanding because many scholars in the humanities focus their efforts on the study of archival materials, which are almost always visual; historians have not yet discovered that it is possible to question the dogma produced by Plato's philosophical tradition. Sound scholar John M. Picker comments on recent critiques of phonograph recordings by Alfred, Lord Tennyson: "It is as if textual critics simply have not known what to do with the poet's own readings and the sound of his voice."[13] William Howland Kenney pins the blame on fascination with the phonograph rather than the voices it reproduces: "cultural analysis of the phonograph and recorded music has languished as writers and scholars alike have favored the study of technology in its many changing forms."[14] We cannot make sense of Tennyson's voice because we have been trained to read visual objects, even technologies, and not sound.

In scholarship, radio seems to have suffered a more tragic fate than the phonograph. Michelle Hilmes writes of radio, "No other medium has been more thoroughly forgotten, by the public, historians, and media scholars alike."[15] In her book about radio, Susan J. Douglas explains the damage caused by our deficit of sonic scholarship: "Existing histories of radio—with the exception of Marshall McLuhan's 1964 best seller *Understanding Media*—do not pause, even for a minute, to meditate on the particular qualities and power of sound, and how these have shaped the power of radio. . . . It is clear that with the introduction of the telephone, the phonograph, and then radio, there was a revolution in our aural environment that prompted a major perceptual and cognitive shift in the country, with a new emphasis on hearing."[16] Voices and noises produce meaning beyond words uttered and recorded, particularly when they are broadcast to millions of listeners. Noises like the ticking of a clock, the dulcet tones of the deep-voiced announcer, and the moaning of the air-raid siren make persuasive arguments to us.

So why did scholars neglect sound for so many years? There are many reasons. Our captivation by visual culture since at least Plato and even more so since the beginning of the age of print has produced a legacy that will take decades if not centuries to overcome.[17] Critics, theorists, historians, and others base their scholarship on what they have been taught, on what has been previously researched, and on available archival resources. Furthermore, they must abide by scholarly conventions that are deeply rooted in science and thus wed to observation and the shackles of visual evidence. Because the voice is ephemeral and fleeting, it cannot be made to fit into the scholarly scientific model that even scholars in the humanities must obey.[18] Rather than cite a conversation with historian Lawrence W. Levine, for example, I am compelled to find related references in his published books and articles. To compound the problem of privileging printed sources as evidence, archives across the world have begun the process of digitizing everything that has ever been published, jotted down, photographed, carved, or painted. Sound has not received the same treatment, and of the few sonic archives that currently exist, many like the private Web site tinfoil.com are almost as ephemeral as speech itself. Thus, a scholar wishing to study sound has more recourse to the printed word than the published song. Scholars also face the problem of reproducing sound in print, a problem that leaves the rare expert to perceive music through visual scores. And a score cannot do justice to a huge variety of noises. Musical as it is, Billie Holiday's voice cannot be heard in quarter notes, glissandos, and annotations. Moreover, sound is much more difficult to reproduce in journals and books than quotes and images. Though some books now appear with companion compact disks, this is not always a feasible option, particularly given the problem and expense of copyright. Academia, its archives, and its publishing conventions are biased toward the visual.

Additionally, until the 1970s, few scholars in the humanities studied the kinds of objects in which sound was most likely to be an emphasis. Because media had long been disdained by "proper" academicians as not worthy of research, popular culture in the form of radio or music was rarely the subject of study. Academics were expected to study symphonies, Shakespeare, and "primitive" folklore rather than rock 'n' roll, *Laugh In*, or dirty jokes.[19] Unfortunately, by the time that popular media became a respectable locus of study, the sonic forms of technology were too old to be of much interest and had been surpassed by film, television, and the Internet in terms of scholarly curiosity. Those who study culture and history are further inhibited by two taboos: that music can only be studied by those trained in musicology and that the study of classical music has now become too elite to be relevant for those who reject the old formalism of humanities scholars during the first half of the twentieth century. Thus, because of his love of Ludwig van Beethoven, some of Theodor W. Adorno's most insightful work is largely ignored.[20]

Adding to the ancient legacies is another more recent predilection: the explosion in the power and primacy of English departments after the turn of the twentieth century dictated to other disciplines in the humanities the importance of the printed word. English departments have trained thoughtful theorists who have propelled scholarship about the word forward at a pace sufficient to maintain the hegemony of this discipline. Scholars of English literature and of the word have focused attention on print, constantly breaking new ground with increasingly insightful theories about how words operate to dominate, hide, and classify all the elements of humanity. Fortunately, some scholars of English have recently begun to rebel against the tyranny of print in favor of studying sound.

Only the field of speech communication maintained the locus for a counternarrative that could have emphasized sound as an object of research. But even scholars in this field gave up the study of the sound of speech after the introduction of television. Taking their cues from theorists trained in English or philosophy like I. A. Richards, Kenneth Burke, and Chaim Perelman, communication scholars abandoned the field of sound criticism. Oddly enough, Burke got his start as a music critic but found that music could only be about nothing; literature on the other hand always had meaning.[21] Like Plato, Burke believed that music most readily produced frustration and fulfillments of desire, emotion rather than logic.[22] To be sure, communication scholars have written about music and about sound in film, but the vast majority of the current corpus is concerned with the lyrics of music or the consumption and production of sound, not sound itself. Lawrence Grossberg, for example, examines popular music for "rock culture," a phenomenon that encompasses the performance and consumption of style, attitude, pleasure, and product but does not critically examine sound itself. Similarly, Stuart Hall's analysis of calypso is a study of identities associated with a form of music but not of sound. These scholars have written important interventions that are to be credited in their own right.[23] Though film, television, and the Internet include a sonic component, scholars have focused almost their entire efforts on the objects of media technology, the visual texts produced by media technology, or the cultural impact of media technology.

So what is the current state of sound studies? The good news is that hundreds of scholars outside of the discipline of musicology have recently begun to devote attention to sound. Indeed, enough scholars have begun to work in the emerging field that some are speaking of a sonic turn.[24] Building on anthropologist Steven Feld's work on acoustic epistemology, the historian Mark M. Smith, for example, has proposed a new field titled *acoustemology*.[25] In the field of media studies, Hilmes has wondered about prospects for the always-emerging field of *sound culture studies*.[26] Similarly, Jonathan Sterne has suggested *sound studies* or *sound culture*, while composer and sound theorist R. Murray Schafer proposed *sonic studies*.[27] The bad news, as evidenced by these five proposed names for one emerging field, is that sound provides such a broad field of analysis in so many disciplines that studies often fail to overlap in terms of vocabulary, methodology, and theory. Historians, literary critics, anthropologists, rhetoricians, and media-studies scholars have sonically examined church bells, ventriloquism, the production of music in South Africa, voices of the dead, the use of sound to sell consumer goods, and Shakespeare's use of shawms. And while the studies just cited succeed marvelously in illustrating something important about a culture or historical period that we can only learn about by reading sound, these studies rarely speak to each other. Sterne's solution is to control this polyglot knowledge by advocating that scholars in sonic studies adhere to principles developed by Karl Marx and followers in the Frankfurt and Birmingham schools like Adorno and Hall.[28] While this is a tempting proposal given the importance of this tradition to the study of sound, Sterne's rejection of phenomenological and psychoanalytical research into sound, for example, would foreclose avenues of study that have also produced important interventions. The solution, I propose, is not to abandon a part of the emerging field that does not make sense to another part of the emerging field but rather to find common ground. Reading sound in the same manner we are trained to read words and images is a good place to start.

But sound is difficult for us to read. Adorno illustrates just how difficult reading sound can be in the absence of sustained scholarship. Though assigned to the Princeton Radio Project after fleeing Nazi Germany, and for decades a scholar of sound, the great theorist, in his *Introduction to the Sociology of Music* (1962), argues that a "good listener . . . hears beyond musical details, makes connections spontaneously, and judges for good reasons."[29] That a good listener judges for good reasons is a tautology, a trap that I have only caught Adorno falling into once. His confusion suggests just how difficult it is to listen, to read, and to translate sound. Adorno struggled for words and even acknowledged that we are all trained to *not* listen.[30] But in the same manner that scholars train themselves to read images, they can also learn to read sound. Few people would have even thought to read images before Michel Foucault's essay on Diego Velázquez' *Las Meninas* (1966),

Michael Baxandale's study on the period eye (1972), or John Berger's book and television series titled *Ways of Seeing* (1972).[31] The accumulation of recent sound scholarship suggests that sonic studies has reached the point that visual studies had reached after these critical interventions.

Sound scholars can easily pick up from not only where visual theorists have taken us but also where sound critics in 1920s and 1930s left off.[32] During the interwar period, two circumstances gave rise to intriguing scholarship that has great potential to further sound studies. On the one hand, the new field of speech communication and the ancient discipline of rhetoric were not yet fettered to Plato or Burke and were thus open to the study of a wide variety of methods, subjects, and objects from speech impediments to theater to classical rhetoric. A read of the contents pages of the *Quarterly Journal of Speech* and its predecessor, the *Quarterly Journal of Speech Education*, provides the twenty-first-century sound scholar with hope. For example, a randomly chosen edition of *QJS* from this period features articles on oratory, "real talk," choral speaking, and curing stuttering, along with reviews of books about applying lessons from the stage to schooling, voice and speech problems, radio speech, and elocution.[33] From this, we can realize that there is much to be learned from studying sound. As A. H. Cantril and G. W. Allport reported in their 1935 study *The Psychology of Radio*, "the constellation of visual cues disappears. Suddenly deprived of the sense of vision, we are forced to grasp both obvious and subtle meanings through our ears alone.... The cues for judging the personality of the speaker and for comprehending his meaning have been immensely reduced."[34] And yet, Cantril and Allport argue that the visually deprived medium of radio improves critical thinking, provokes curiosity, and strengthens democracy.[35] Following in this tradition, Joshua Gunn argues that there is "something more in speech than speech."[36] For Gunn, speech is magical, dangerous, and important; thus, he has called on scholars to reinsert sound into the study of communication.[37]

Adorno anticipated Gunn in his belief that the close analysis of sound was critical to efforts to better understand history and culture. Known now as a cultural critic in general, during the 1930s and 1940s, Adorno published scores of articles and books about sound, ranging from essays about classical music, to radio audiences, to religious broadcasters. Writing of the importance of studying the sound created by his favorite composer, Adorno argues, "If we listen to Beethoven and do not hear anything of the revolutionary bourgeoisie—not the echo of its slogans, the need to realize them, the cry for that totality in which reason and freedom are to have their warrant—we understand Beethoven no better than does one who cannot follow the purely musical content of his pieces, the inner history that happens to their themes. ... If we wished to catch up, to release the cognition of music from its inane isolation, it would be necessary to develop a physiognomics of the types of musical expression."[38] For Adorno, to listen critically was to recover the past. Beethoven's music is not just pleasing; it is a historical text that permits us to understand the growth of enlightenment and the revolutionary urge for modernity. Music should not be isolated but rather incorporated into fields like history and anthropology through theoretical frameworks like physiognomy. Though Adorno was still partly beholden to visual texts (he read the notations of musical scores, for example), he believed that sound could and should be read of its own accord. Interestingly, Adorno references the discredited theory of physiognomy by which the physical features of an individual could be read to determine characteristics like kindness, madness, and criminality. Pseudoscientists employing this method argued, for example, that a large forehead was a mark of intelligence or a diminished chin the mark of ethical weakness. Adorno employs physiognomy to metaphorically argue that musical passages, a sforzando or crescendo, for example, can be read to illustrate larger cultural phenomena like the desire for change or the culmination of revolution.

Adorno's pairing of sound with a discredited visual theory is intriguing. Sonic scholars often find themselves in opposition to visual theorists who privilege image over sound. Douglas, for example, argues that "with all the academic attention on the power of 'the gaze,' the power of hearing to shape individual and collective subjectivity has gotten short shrift. It's time to rectify this."[39] Theories about the gaze by Jacques Lacan, Foucault, and Laura Mulvey provide extraordinary insights into how humans create selfhood and identity, the surveillance society, the categorization of otherness, and the dominance of the masculine eye in cinema even after second-wave feminism.[40] For each theorist, vision dissects and provides the critical tool for analysis. While the intellectual descendents of Lacan, Foucault, and Mulvey continue to privilege the gaze over other forms of perception, there is no reason why their scholarship cannot be repurposed to the study of sound. Indeed, hearing practices can be tied to visual practices.[41] Not surprising, psychologists have two terms for the conflation of vision and hearing. The first term, *synesthesia*, applies to the overlapping of any pair of senses. One may feel something that is otherwise a visual phenomena, as when one experiences "the blues." Or one may taste something that is otherwise felt as when one eats "sharp" cheddar cheese. The conflation of senses is also found in musical notations like # (sharp), phrases like "loud shirt" and high or low notes, and an angled symphonic passage. The most common form of synesthesia is the conflation of sound and color and has its own term: *pseudochromaesthesia*.[42] This conflation appears in the musical genre called "the blues," in "white noise," and in the perception of an ascending trumpet blare as "bright." The vestigial remains of words sometimes illustrate the phenomenon. When we speak of the musical flourish that introduces a head of state, for example, we recall the colorful flowers from which the word is derived.

Throughout this book, I turn to the work of visual scholars to make sense of sound. This by no means limits the avenues available to other scholars wishing to study the intersection of sound and criticism. Employing analogies between visual and sonic criticism is particularly useful to my project for two reasons. First, there is a historical parallel between the development of visual studies beginning in the 1970s and the development of sonic studies during the first decade of the twenty-first century. Second, the very terms employed by scholars in the first half of the twentieth century when the importance of sonic rhetoric was most obvious provide modern scholars with a visual vocabulary for studying sound. Adorno was not alone in using a visual term to understand the sonic. Furthermore, recent studies in visual rhetoric provide a number of useful methods, terms, and concepts for the study of sonic rhetoric. Visual scholars have provided a vocabulary for understanding image vernaculars and period eyes—manners of perception unique to a culture and era. Applying this scholarship to an essay about the period ear, Shai Burstyn notes the importance of cultural assumptions and the listener's role in making sound meaningful.[43] During the presidential election of 1932, one gets a sense of how popular and unpopular the two candidates were by understanding this period ear: "Voices 'paint' character on the radio. . . . The mere sound of a voice, the intonation, inflection and power of delivery can portray sincerity, cheerfulness, self-confidence, shrewdness and other qualities. . . . President Hoover's voice betrays deliberate effort, according to John Carlisle, production manager of the Columbia Broadcasting System, who labels the Hoover voice 'typical of the engineer.' He calls Governor Roosevelt's voice 'one of the finest on the radio, carrying a tone of perfect sincerity and pleasing inflection.'"[44] Carlisle, a radio production manager, applies stereotypes to describe the extraordinarily unpopular Herbert Hoover, a president who had presided over Black Tuesday and the beginning of the Great Depression as sonically deliberate and cold like the engineer that he was. Roosevelt, on the other hand, broadcast in a tone that was fine and pleasing in a manner that reflected his perceived sincerity. That the tone of each man's spoken words was interpreted to mean what audiences expected to hear should not surprise us because the voice has long been regarded as either the soul itself or a marker of the soul.⁴⁵ Furthermore, not only does the period ear guide interpretations but assumptions about sound are also resources that we draw upon to construct arguments that make sense of our cultures and of the visual and sonic phenomena that we witness. In 1932, millions of Americans employed their assumptions about the sincerity of Franklin Delano Roosevelt's voice to persuade themselves into voting for him. Reading sound enables scholars to understand why.

Similar to the period eye, scholarship about visual icons can be applied to sonic icons. Where rhetorical scholars Robert Hariman and John Lucaites study images like "Migrant Mother" and the raising of the flag at Iwo Jima that help to shape and inform citizenship in the United States, certain sounds may accomplish similar tasks. And where Hariman and Lucaites find their visual images repurposed as political cartoons and advertising gimmicks, I find iconic sounds like the famous series of notes in Beethoven's Fifth Symphony to be markers of an historical era and persuasive objects that have been adapted to sell movies and automobiles. Similarly, Road Runner and Wile E. Coyote cartoons have perpetuated the now-iconic sound of a falling bomb. Like the images that teach us how to be citizens by reminding us of sacrifice and heroism, the sounds remind us not to be the people being bombed. We should not be surprised at all that NBC's tritone and Intel's four-note tag are as recognizable as the NBC peacock and Intel's blue orbit.⁴⁶

Lastly, visual scholars have produced standalone arguments about images. Most famous, Berger includes an image-only essay in his book *Ways of Seeing*. I am excited by the sonic possibilities raised by this intervention. Perhaps, sounds are also preverbal and do not need words to make sense of them. Rhetorician Kevin Michael DeLuca's call to shift critical attention from the gaze (which assigns meaning to visual phenomena) to the glance (which assigns agency to visual phenomena) may similarly be applied to listening and hearing.⁴⁷ Like those who spend their lives uncritically being manipulated by images, few people critically listen to objects. Instead, we spend our lives immersed in sound but rarely do anything more than hear it. Additional potential is in visual theorist James Elkins's book *The Object Stares Back*.⁴⁸ If objects stare back, then objects must also listen back and in doing so inform us of who we are by creating for us identity, experience, perspective, and context.⁴⁹

Although visual scholars have provided me with a useful, if metaphorical, vocabulary for the study of sound, elsewhere I fall back on my training as a reader of words and thus employ long-established techniques like thick description and close reading. Inevitably, I am a creature of the system that produced me, a relentlessly visual system that has deprived me of the critical use of four of my senses. Theoretically, I cannot wander far from my intellectual heritage, and so I find comfort in adhering to it when I am most confused. As a result, I have necessarily excluded from this study sounds that I am not yet capable of reading. I had hoped, for example, to read a *phonautographe* that was recorded by the French inventor Édouard-Léon Scott de Martinville in 1860. The inscription of sound into image, for that is what his recording is, was produced by Scott's machine and, likely, his own voice. It was never intended to be played back, and until 2008 no one could have imagined listening to his etching. But thanks to modern inventions and in particular to a recently discovered ability to digitally translate physical objects into sound, we can hear again Scott singing a passage from a French folk song. Although it is perfectly feasible to situate the sound within the context of technology in 1860, the genealogy of Scott, and the cultural, intellectual, political, and artistic atmosphere during France's Second Empire, I am not able to read the sound itself. Would that this was the only sound I was unable to read.

I also take comfort in a favorite passage penned by Ralph Waldo Emerson: "A foolish consistency is the hobgoblin of little minds." Hence, I feel no insecurity at being able to read some sounds but not others; sound after all is hard to read. Similarly, I feel no compulsion to bend sound waves to one particular manner of reading sound or one specific theory. Thus, I employ theorists from a variety of traditions and periods—Adorno, Berger, Pierre Bourdieu, Lacan, Walter J. Ong—without concern for contradictions. I feel a need only to suggest the reasonableness of reading sound and in so doing provide evidence that reading sound can be done using a variety of alternative and sometimes contradictory approaches. The viability of a specific method for reading sound is not as important as the greater argument that sound can be read.

Lastly, this study is in many respects an effort at synthesis though not of history or theory but rather of sonic criticism. Detractors may argue that the book covers too much ground or that it fails to dig deep enough. I would agree with both critiques if this book was about the sound of presidential oratory, or the 1950s, or psychoanalysis, but it is not. It is a book about the practice of reading sound with illustrations that are, by and large, chronologically organized. I leave historical and theoretical studies to other scholars. As Levine was fond of telling his graduate students, there is a need for books and essays that dig to the very root of a historical, rhetorical, or theoretical question. These monographs enable scholars to break new ground and to claim a kind of truth. But Levine also encouraged works of synthesis. He believed that these would either summarize the past, thus enabling scholars to reach a wide audience, or would spur new research in reaction to the synthesis, creating new relationships and spurring new scholarly conversations. A book devoted to sonic criticism and its relation to other forms of scholarship in the humanities may help future researchers to discover new methods for interpreting their primary sources and thus enable them to dig even deeper into meaning, identity, culture, rhetoric, and history. Research into sound may also provoke a rethinking of visual scholarship or identity politics and may encourage more academics to further study the history, psychology, and culture of the five senses or may provoke conversations among scholars from different traditions in the humanities. For now, I only hope to show that sound can be read and that Levine was right: we can listen to our sources.

Organization

This is a book about reading sound. The chapters exhibit different methods of close listening and employ dissimilar sets of theoretical perspectives, beginning with rather simple premises and ending with more-complex interventions. I do not strive for a consistent methodology here, preferring instead to constantly reiterate my thesis that sound can be read and, what is more, read in a variety of ways—just as images can be read through a variety of perspectives. Though I hope that the inchoate field of sonic studies manages to find methodological common ground, here I only posit the most basic of universal tenets: that we can find common ground through the critical analysis of sound.

Chapter 2, "Fitting Sounds," begins with the simple kind of sonic reading that we are taught from an early age to do: to read for accent and dialect. All people demarcate identity boundaries, not only through visual cues like skin color and style of dress but also through audio cues, the most important of which are the foreign accent and the regional or class dialect. After reading for accents made on a pornographic-recording cylinder of 1895, I develop evidence of a change in the sound of the presidential voice during the period from 1892 to 1912. Perceptions of and expectations about the presidency shifted during this period as evidenced by visual clues (political cartoons) and by verbal cues (which words were used). Moreover, the very sound of the presidency changed. Working through primary documents from novels to public-speaking textbooks to the reminiscences of actors, I illustrate that a significant shift in the sound of oratory occurred during this period. Throughout, I also read the sound of recorded presidential speeches, culminating with "The Right of the People to Rule," an oration by Theodore Roosevelt during the campaign of 1912. Indeed, the people ruled Roosevelt's speaking style; he was the first recorded president to sound like the people.

Chapter 3, "Machine Mouth," examines the impact of modernity on listening patterns. Here the focus is on two sounds that are iconically modern, the clock and the locomotive, and yet were bothersome. The trouble with adapting to modern noise is exemplified by Richard Wright's novel *Native Son*, a novel whose protagonist fights with and loses to the sound of modern time. The war of the working class against the clock appears elsewhere in literature and in film, backing up Wright's semiautobiographical frustrations with modernity. Of course, clocks were not the only alarming feature of the modern soundscape. Other novelists and filmmakers complained of the endless humming, pumping, and blaring noises of all machines. But hypermodern composers like Igor Stravinsky and Arnold Schoenberg and painters like Pablo Picasso and Umberto Boccioni celebrated and adopted the penetrating noises, turning these into self-shattering concerts and works of visual art. The dream of adopting and adapting to modern noise was shattered, however, by the cacophony of a Great War waged between 1914 and 1918. As a result, a new interpretation of modern noise emerged as epitomized by the bending of sound into curvilinear art forms and its assimilation into the popular consciousness through music. Hitler exploited the shattering noises of warfare and machinery by enveloping his listeners in sound through microphones, loudspeakers, and radios with disastrous consequences. In America, musical artists comforted listeners by training people like Wright to adapt to the noises of modernity. Booker "Bukka" White's tune "Special Streamline" exemplifies this effort.

Chapter 4, "The Race of Sound," examines the use of sonic tropes as they were applied to race during the interwar period. This chapter features close readings of a recording of an ex-slave named Phoebe Boyd made by members of the Dialect Society of America in 1935, a 1936 radio episode of *Amos 'n' Andy* (a program that featured white actors sounding "black"), and Holiday's song "Strange Fruit" (1939). These recordings illustrate the use of sound to create racial separation and, ultimately, identity. During a period when concerns about race pervaded the American consciousness, listeners often had to make sense of identity without visual cues. As a result, assumptions and expectations about the sound of races came to dominate the period ear. Not surprising, then, "race records" appeared in 1922 and the phrase *black music* in 1924. While these terms may seem odd to present-day readers, psychological accounts from the period illustrate a belief that the sound of one's voice was a true reflection of the individual's self and, in particular, a person's race. Thus, "black" and "white" voices were more difficult to integrate than black and white bodies.

Chapter 5, "Sounds of War," describes the use of sonic manipulation by government agencies and Hollywood producers to affect and contest public opinion. Edward R. Murrow's recordings of civil-defense sirens over London, for example, helped persuade Americans to join the Second World War on the side of the British. During the 1950s, *Duck and Cover* and other educational films employed the same sirens to cast fear into the hearts of Americans. But Hollywood contested the sonic manipulation by the military-industrial complex. Most famous, Chuck Jones's Road Runner cartoons (1949–66) featured a running gag with a perpetually falling Wile E. Coyote, who reproduces the sound of falling bombs to great comedic effect. Conversely, President Lyndon Baines Johnson employed a powerful combination of sound and images of nuclear Armageddon in his 1964 Daisy ad. These sonic manipulations continue—in Hollywood films, political advertisements, and Pentagon propaganda—and their persuasive impact has only increased.

The sixth chapter, "On Sound Criticism," is an epilogue that is designed as a complement to Mark Smith's final chapter in *Listening to Nineteenth-century America*. In that chapter, Smith's essay on method describes the emerging field of sonic studies as it existed in 2001. A decade later, I happily have more to work with and thus more ground to cover. The epilogue, then, serves as a summary both of this volume and of scholarly efforts to read sound. It is more a meeting point than a starting point for theoretical perspectives about sonic criticism. Given all the work that has been done in the last decade and given the recovery of older scholarship, it is a hopeful chapter that makes real Levine's directive to listen to historical sources. Although there are many vocabularies, theories, and methods, there is also common ground and in particular an impulse among scholars in the humanities to read sound as they have been trained to read print and images. This book will not be the last word about reading sound. Other voices will inform scholars in innovative ways that sound can be read. I look forward to listening.

2

Fitting Sounds

During the 1890s, millions of Americans tried out a surprising invention produced in Thomas A. Edison's Menlo Park factories. Eleven years after its invention in 1877, Edison had finally made the new device marketable. The automatic phonograph, a "nickel-slot" machine (essentially, a jukebox with one recording) that salesmen introduced to saloons, train stations, and other crowded locations, played back prerecorded cylinders for five cents. Most of these cylinders carried the sounds of storytellers repeating jokes, minstrels engaged in slapstick comedy, and popular and classical music. A few of these recordings were political speeches, and a few others were pornographic.[1] Plunking down a nickel, listeners could enjoy the sounds of an entertainer swearing repeatedly or telling a dirty joke. Some of these cylinders were so pornographic that one performer, Russell Hunting, was imprisoned in 1896 under the Comstock laws for the suppression of vice.[2] Hunting was one of the first stars of the media age, known especially for his Michael Casey series of humorous recordings in which the actor imitated an ignorant Irish immigrant.

Before he was arrested, Hunting produced "Out of Order."[3] The recording, one of the few obscene cylinders from the 1890s that can still be played, is remarkable for a variety of reasons. Obviously, Hunting's performance disturbs any notion we may have had about the moral purity of early sound recordings. Though we know that minstrel-show cylinders were racially and ethnically offensive and often misogynistic, these do not reach the level of "I'll know it when I hear it" pornography attained by Hunting's recordings. We may be surprised to hear that Victorian-era Americans willingly dropped nickels into a machine to listen to a three-minute recording that employed foul epithets or were hateful toward women. But if we know that readers paid for smut in books and that friends told each other dirty jokes, then we should not be surprised that "obscene, lewd, and/or lascivious" materials made their way onto sound recordings during the earliest days of the phonograph's history.[4]

What is particularly striking about the cylinder titled "Out of Order" is its sound. The original recording is hard to hear. Playing the cleaned-up and digitized version of the cylinder points out the difference between passive hearing and active listening. To understand the sound of this recording, one must pay very close attention to the words. Though twenty-first-century sound engineers have succeeded in filtering out some of the clicks, pops, scratches, and background fuzz that pervade all early recordings, the voices still remain more difficult to decipher than what digital-sound consumers are accustomed to hearing. For example, the voices often clip (the noise exceeds the limits of the recording and playback equipment and thus emerges distorted) and are what has been appropriately described as tinny, or in other words having a metallic sound as if the voices emerge out of one of two tin cans attached by a string. Scholars might insist that listeners in the 1890s had the aural skills necessary to interpret the noises and voices with ease; however, this technology was new, and most listeners were only able to hear the three- or four-minute recordings at crowded and noisy saloons, train stations, and county fairs. In the 1890s, listeners had to pay very close attention to understand.

"Out of Order" is also remarkable because it features a studio audience. Radio introduced studio audiences in the 1930s so that comedians would know if their jokes were hitting the mark—thus indicating whether the delivery should be changed—and, more important, so that home audiences would feel comfortable laughing at a machine. In the 1950s, television perfected the manipulation with the laugh track. Already in the 1890s, phonograph companies were employing similar devices. Given that Hunting would not have needed to know if he had hit the mark because no further dialogue occurs after the punch line, the recording's producer must have intended the studio audience to cue nickel-slot listeners to laugh at their automatic phonographs.[5] The laughter at the end of the track informs listeners that they are part of a community, an effect that meant the audience is doubly in on the joke—they get the joke, and they are members of a clique that has forbidden knowledge.

At the heart of this recording is the mixing of dialects and accents in late nineteenth-century America. The dialogue has five parts, each with its own accent or dialect and each performed by Hunting. The first is the announcer, typical for recordings in the 1890s, who introduces the cylinder as a "humorous record by Charlie Smith of Kankakee." The announcer speaks in a light version of the era's orotund intonation. He emphasizes the letters *m* and *t* in words like meeting, committee, and countryman and rolls his *r*'s in *humorous, three,* and *were*. Even so, most of the *r*'s are not trilled, and some phrases like "scene, uh, illustration" are intentionally sloppy. Indeed, Hunting employs many *uh*s (seventeen in one minute and seventeen seconds) to break up the formality of the introduction.[6] Furthermore, he employs colloquial expressions like "kicking," "go-by," and "freeze him out." Although most recordings of this decade employed introductory announcements that were enunciated in theatrical English, this recording intentionally violates the norm. Not surprising, the announcer also rejects the norm of broadcasting the name of the company responsible for the cylinder. The phonograph's maker might have suffered the consequences of the Comstock laws had the affiliation been known.

The next part of the recording is the local chairman of a bureaucratic arm of Bloomfield. He has been charged with overseeing a committee tasked with locating a water reservoir. Hunting changes his voice, making the tone deeper and the amplitude louder, perhaps by way of mimicking the power of any chairman. Hunting's chairman occasionally apes the rolling *r*'s and hyperenunciation of orotund speakers and generally sounds formal, stiff, and humorless, his pronunciation flawless but for a few *uh*s. Most listeners in the 1890s would have perceived the chairman as a white male and as someone who sounded educated, authoritative, and intimidating. Indeed, even in the twenty-first century, the chairman sounds bureaucratic.

The third voice, a countryman named Mr. Wilkins, speaks in a backcountry dialect that emerges in a nasally twang. This is the voice of a rube. The baritone voice of the chairman contrasts with the countertenor (i.e., falsely high) voice of Wilkins, thus depicting a hierarchy of sonic authority: the lower the voice, the more masculine and authoritative. Hunting also slurs Wilkins's speech, running words into each other, dropping consonants, and employing slang like "t's" instead of the proper "it is." Like the chair, Mr. Wilkins would have been perceived as a white male, though with little or no education. American schoolteachers in the nineteenth century trained children not only in proper grammar but also in proper speech. Mr. Wilkins's evinces little sign of any such training.

A fourth and nameless speaker has an obvious German accent and is portrayed by Hunting bastardizing words like "was" to "vas," "that" to "dat," "and" to "und," "whatever" to "vatever," and "farmer Wilkins" to "farmer Vilkins," giving away the occupation of the third speaker. Hunting returns this character to the same pitch that he used for the introduction because he knows that the tenor voice of the immigrant does not pose a threat to the authority of the announcer. We are beginning to hear a complex sonic hierarchy in the late nineteenth century. "American" voices move along a vertical axis from sopranos with no authority (like women) to basses, who exemplify authority like the mythologized voice of Abraham Lincoln. Voices with foreign accents can move down the sonic scale toward a deeper voice without increasing their authority; the voices are still just those of powerless immigrants, after all. The German and the announcer have the same pitch in this recording, but most listeners would have interpreted the announcer as having authority because he spoke without a foreign accent.

The last voice is that of an Irishman, Reilly, with a thick accent and a tenor voice again. Reilly swears like the proverbial sailor, employing *begod* and *damn* without hesitation and butchering words with the brogue accent that elite Americans had condemned for much of the nineteenth century. Interestingly, Reilly opens with a phrase employed by the announcer, "I was froze out of it," and is brought up short by the chairman, who demands an explanation of this slang. This colloquialism identifies Reilly with listeners who do not need the phrase explained (otherwise, it would have been clarified after it had first been uttered in the introduction) and separates the chairman's formal education from the vernacular knowledge of Reilly and the audience.

The joke turns on the difference between the vernacular-literal and educated-figural interpretations of words. The Irishman complains that the proposed reservoir was so small that he could "piss halfway across it, sir." This brings a rebuke from the genteel chairman: "Mister Reilly, you're out of order." Here, the audience switches back to identifying with the judge and interprets "out of order" in an educated manner, recognizing the phrase's use in bureaucratic situations. In the hands of a judge or a chairman, calling someone out of order is a reprimand for speaking out of turn or for saying something inappropriate or indecent. But the uneducated Reilly takes the judge literally and concludes that the judge has questioned his manhood. Responding with a stream of swearwords, Reilly exclaims, "Well, begod, sir, if I was in order, I could piss the whole damn way across it, sir!"

That Reilly employs the term *sir* twice in the sentence reminds us that he is not trying to offend. Rather, he is employing the proper form of address and remains deferential, understanding that the chairman is the authoritative figure in this sound play. Reilly even manages to pronounce *order* with the chairman's inflection, including a lightly trilled *r*. Accidentally, Reilly challenges the rules of the language game through his ignorance. He does not know that he should not swear in front of a chairman, so much so that even after he's been rebuked, he swears three times. But the joke works, then and now, because the audience knows the rule and yet participates in the attack on the formal and intimidating bureaucrat by paying to listen to the breaking of the rules of official discourse.

What is particularly remarkable about the recording is that the broguist has a name and that the chairman recognizes the manhood of the speaker: "Mr. Reilly, you're out of order." Reilly is a citizen and thus commands respect even from an educated chairman. Rather than entirely excluding Irish brogue from bureaucratic institutions that projected an official version of a purified English, the chairman respects Mr. Reilly.[7] And even as audiences laugh at the manner in which rubes and immigrants speak, they participate in the play of the rubes' language games, challenging bureaucratic-speak and elite intonation.[8] Both sides are the butt of the joke, the Irishman because he takes "out of order" literally and the chairman because he has allowed the Irishman to mock officialdom and its patterns of speech. The rube may be stupid, but the bureaucrat is silly for trying to impose an official language: "out of order" really does also mean not working properly.

Moreover, the consumption of a recording that features the dialect and accents of an elite chairman, a country rube, a German immigrant, and an Irish-born citizen indicates that for the class of listeners who plunked their nickels into the automatic phonograph's slot, diversity of speech was tolerable though worthy of being joked about. The accents are a parody, yet their place in the mechanisms of government are not unspeakable. Indeed, at least some of the consumers of this joke spoke like the satirized voices. While the number of dirty jokes recorded onto cylinders employed for public consumption may have been small (twelve such recordings are known), the number of recorded jokes employing rube, city slicker, black, Irish, German, or Yiddish accents is great.[9] Beyond the sheer number of recordings, their penetration was also significant. Phonograph makers and distributors made huge profits off of the nickel-slot machines as millions of Americans listened in on an always-changing assortment of musical and comedic recordings. Comedians like Hunting became stars among nickel-paying laborers and, increasingly, middle-class and elite families able to purchase their own phonograph players. William McKinley and Theodore Roosevelt, for example, both owned phonographs and "special" (i.e., pornographic according to nineteenth-century standards) cylinders.[10] Though these two presidents may have surrounded themselves with like-minded people who spoke in the same dialect, the two presidents listened to jokes like "Out of Order" in which government bureaucrats entertained the complaints of men with accents and were simultaneously made fun of for their pretentious style of speaking. The recordings illustrate that the definition of *American* was changing in a manner that incorporated those who did not speak "proper" English.

Americans of all kinds were faced with the constant presence of accents and dialects, in recordings and in person, and in conversation and in jokes told around the kitchen table. By 1900, 13.5 percent of all Americans were not native English speakers.[11] Adding to the mix of Chinese, German, Italian, Slavic, and Irish immigrants all trying to get by in English came the first wave of black migrants from the south, who introduced their own dialects into the cultural production centers of New York, Boston, Chicago, Philadelphia, and Washington, D.C. And while these linguistic intrusions into the wealthy, white, Northeastern dialect are evident, yet another intrusion was also occurring. The growing power of the labor movement and the organized farm movement in America brought the working-class voices of the steelworker and rancher into the boardrooms of corporate America and the smoke-filled backrooms of the federal and state capitols.[12] Few could have avoided the mix of accents and dialects in late nineteenth-century America. Not surprisingly, that mixing of accents produced a dramatic change in the American dialect.

The most dramatic of these turn of the twentieth-century changes occurred in the sound of the man who was supposed to represent all Americans: the President of the United States. With the succession of Theodore Roosevelt to the presidency after the assassination of McKinley, that change was both visually and sonically apparent. The svelte, mustached Roosevelt looked nothing like the hefty, clean-shaven McKinley. The appearance of fitness for the presidency after the turn of the twentieth century required not only the weighty old political *gravitas* but also a new manifestation of manliness as well. Roosevelt defined this fitness in his book *The Strenuous Life* (1899) as "the life of toil and effort, of labor and strife . . . the man who does not shrink from danger, from hardship, or from bitter toil."[13] Thus with steel will, Roosevelt visually appeared to be prepared at a moment's notice to work hard for Americans and to face the dangers of a modernizing world. The two presidents sounded different, too. In public, McKinley took on the stentorian tones of the Shakespearian actor, a style that we today would consider bombastic and pretentious but that audiences expected during the late nineteenth century. Indeed, public-school speech teachers taught this style. Roosevelt, on the other hand, sounds like a professor in recordings made of his voice. Turn-of-the-twentieth-century Americans demanded and witnessed a dual revolution, visual and sonic, in their expectations about the presidency.

Dialect and Manliness

Sound can be read, if for nothing else, for the interaction of dialects and their contextualization. While a transcription of Hunting's "Out of Order" might permit readers to catch accents, a printed account would not produce as deep an understanding of expectations about authority, nor would we be aware of how closely listeners "read" the voices broadcast by automatic phonograph machines. New technologies and a multitude of accents would have forced audiences to pay attention, lest their hard-earned nickels be spent in vain. Though variations in the sounds of American English would remain contested well into the twentieth century, the comedians in the late nineteenth century who poked fun at elite oratory, rube dialects, black vernaculars, and immigrant accents were setting the temper for what would become a sonic language-of-state that is exemplified today in the voices of news anchors. And while these exemplars have been condemned for homogenizing the sound of American English into a universal, "white bread" dialect, at least they do not sound like the elitist orators of the late nineteenth century. Even the elites had to bend to a popular-language vernacular.

At least three reasons explain why popular culture influenced presidential oration and the relationship between oratory and manliness at the turn of the twentieth century. Significant social change during this era—in particular, immigration, populism, and new recording technologies—influenced rhetorical culture, which caused the sound of the presidency to change. Lawrence W. Levine explains that in the first manifestation of popular culture's power, oratory changed to account for millions of new immigrants.[14] If authorities were to persuade, they had to speak in terms that immigrants could understand. Simultaneously, immigrants threatened to undermine the definition of *American*. Cara Finnegan makes a similar turn to immigration in her explanation for odd readings of a recovered image of Abraham Lincoln. In 1895, *McClure's* magazine published a rediscovered daguerreotype of Lincoln and others' reactions to the photograph that tell little about the image but do inform about visual expectations. Late nineteenth-century Americans were taught to see leadership, intelligence, and other traits in the visible features of their presidents. Thus, viewers saw in Lincoln's rediscovered image an ideal American type. This vision suggests, for Finnegan, that white Anglo-Saxons were struggling to maintain control over the meaning of *American* at a time when millions of immigrants were entering the United States, a problem resolved by promoting a tall, white, male exemplar who had been a country rube and yet was sonically mythologized as a blue-blooded Boston Brahmin.[15] Immigrants brought visual pressures to the definition of *American*. They also brought sonic pressures.

Beyond immigrants, the laboring classes also increasingly influenced both the appearance and sound of presidents. By 1900, corporate trusts had failed to bust the unions, and Samuel Gompers's American Federation of Labor exercised growing political clout.[16] Combined with powerful agrarian organizations, laborers and farmers assailed the fat crabs (*fat crab* was later replaced by *fat cat*) who controlled the levers of power through privilege and wealth. Thus, when the comparatively svelte William Jennings Bryan traveled across the country during the elections of 1896, 1900, and 1908, he spoke to the people—laborers, farmers, clerks—with a manly appearance that mirrored and a voice that echoed their own. Not surprising, he was nicknamed "the great commoner." That he did not win the first two elections against a heavy-set candidate who spoke in the orotund style, discussed later, testifies to the lack of penetration of new media technologies and not to Bryan's oratorical abilities. Unfortunately for Bryan, only a small number of Americans were able to hear him speak in 1896 and 1900. Similarly unfortunately, by the time that sonic media had become common, Bryan's last opponent had also switched to the common style of speaking.

It was not until the elections of 1908 and 1912 that phonograph recordings of political speeches became widely disseminated, and this is the third reason I propose for why the sound of the presidency changed.[17] Thanks to many innovations made to the phonograph, by 1908 that technology had become common in the saloons of working-class Americans and the parlors of middle-class Americans.[18] Campaign disks and cylinders were sold in 1908 by the Victor, Columbia, and National labels, which all invested resources into producing and marketing dozens of these recordings.[19] Presented with a mass audience of listeners who represented every class and dialect in the spectrum, presidential candidates could no longer orate in the orotund manner. They had to sound like the voters, who were now listening, and in a manner appropriate to venues, like saloons and parlors, where the orations were heard.

While these changes recomposed the rhetorical culture of the early twentieth century, specifically moving political oratory toward the common style, two parallel shifts contributed to the emergence of a new kind of manly speech: the rise of working-class manhood and the demise of middle-class virility. The French sociologist Pierre Bourdieu finds contests over manly talk in his study about the imposition of "official language," which is the formal and imagined national dialect. Bourdieu argues that modern states employ educational systems and bureaucracies to impose a language purified of regionalisms, accents, and dialect variants.[20] Yet, he also maintains, official languages remain contested along regional and class lines.[21] Bourdieu comments, for example, that the contest between political elites and the working class over dialects produced condescension: "[T]he strategy of condescension consists in deriving profit from the objective relation of power between the languages that confront one another in practice . . . in the very act of symbolically negating that relation, namely, the hierarchy of the languages and of those who speak them."[22] Normally, this enables the respected speaker to talk in the people's style, which conveys authority and reifies the speaker's cultural capital. He or she only condescends to speak like commoners even as the people recognize the orator normally speaks in an official dialect. The strategy of condescension is most conspicuous during election campaigns when candidates use the common style to appeal to provincial voters. Through the use of the appropriate dialect, politicians imply by their performance that they are like their listeners and thus share the same values and political opinions.[23] William Jefferson Clinton and George W. Bush, politicians trained at Harvard and Yale universities, speak with southern accents. At the turn of the previous century, candidates for president trained at Yale (William Howard Taft) and Harvard (Theodore Roosevelt) similarly spoke to the people in the common style. This style has been perceived for over a century as appropriately American and manly.

Occasionally, however, a particularly elite kind of condescension is instead interpreted as "the hyper-correction strategies of pretentious outsiders, who are thrown into self-doubt about the rule and the right way to conform to it, paralyzed by a reflexiveness which is the opposite of ease, and left 'without a leg to stand on.'"[24] It is this phenomenon—overly official speech—that may have propelled Roosevelt to speak in public like the people rather than like exemplars from his elite heritage. After Roosevelt, orotund style goes quiet, except in farce. Not surprising, when Bourdieu turns his focus toward responses to the imposition of official language on the working class, he finds that laborers sometimes resist by mocking the formal style as effeminate and by adopting nonconforming dialects and the common style.[25]

In 1905, Tammany Hall politician George Washington Plunkitt exemplified the working-class reaction to pretentious talk when he advised that orotund speech would only win votes in districts dominated by the effeminate. Hinting to one aspiring politician who "was beginnin' to put on style," Plunkitt warned, "The Democrats of this district ain't used to dukes and princes and we wouldn't feel comfortable in your company. You'd overpower us. You had better move up to the Nineteenth or Twenty-seventh District, and hang a silk stocking on your door."[26] Equating elites with effeminacy was a bold statement.[27] Plunkitt continued, "Another thing that people won't stand for is showin' off your learnin.' That's just puttin' on style in another way. If you're makin' speeches in a campaign, talk the language the people talk. Don't try to show how the situation is by quoting Shakspere [sic]."[28] Rather than confront immigrants and laborers with the orotund style, Bryan, Roosevelt, Taft, and Woodrow Wilson all adopted a common style of speaking. Plunkitt's admonition illustrates both the growing power of the working class and a rejection of what they deemed to be effete political culture.

A second reason for the emergence of a new manly style can be found in the concerns about the supposed feminization of sedentary white men. Between 1870 and 1910, the number of these men who worked in "white-collar jobs" (a term coined in 1919) increased eightfold.[29] Because they no longer used their muscles, they became concerned about their own manliness, as well as representations of that manliness in their elected representatives.[30] Roosevelt's penchant for backcountry travels and big-game hunting exemplifies this concern. He wrote in the *New York Tribune,* for example, "For good healthy exercise I would strongly recommend some of our gilded youth go West and try a short course of riding bucking ponies, and assist at the branding of a lot of Texas steers."[31] *Gilded* youth branding *steers*? Roosevelt here seems insecure about castration and the loss of white male virility: *gilded* is a near homonym of gelded, and *steer* is a castrated bull. Roosevelt wanted people to hear him, through even the printed word, as if he talked like they did (and recall that plenty of Americans were not far removed from illiteracy and speaking words as they read). The homonyms created what rhetoricians call "productive ambiguity." Furthermore, the words look similar. The rejection of elite political culture as effete and the widespread concerns about the loss of middle-class American manliness were linked to each other through popular productions like *Scouting for Boys* and *Algie the Miner,* not to mention Roosevelt's autobiographies like *Ranch Life and Hunting Trail* and *The Rough Riders.*[32] American men, and therefore voters, were concerned about their manliness and the virility of their representatives. If the voters themselves could not be manly, at least they could idealize their manliness through elected representatives who exemplified virility. Given the influx of foreign-speaking immigrants, the increasing political clout of labor, the rapid dissemination of recording technologies, and the concerns about virility, it is not difficult to hear how the sound of the presidency irrevocably changed from the orotund style to a style that echoed how laborers and the middle class spoke during the first decade of the twentieth century.

The Orotund Style

Grover Cleveland wore his weight like the *gravitas* it represented in the 1880s and 1890s. But Taft had the gross misfortune of being fat at an age when Americans no longer viewed girth as a marker of power, a reminder of how quickly cultures can change. For many industrializing nations, fat becomes a signifier of one's ability to consume and thus one's power. But even as fat became a signifier of power in late nineteenth-century America, it also became a visible trope that contrarian cartoonists could employ to exaggerate cultural assumptions. Not surprising, these cartoonists learned to apply such an exaggeration to those who had wealth and power. Thus, trusts like John D. Rockefeller's Standard Oil came to be represented as incredibly large, so much so that they not only outweighed the people they dominated but they also lorded over their subjects.

In Joseph Keppler Sr.'s cartoon, "The Bosses of the Senate" (see fig. 2.1), thirteen of these massive creatures have entered the monopolists' entrance to the U.S. Senate and now overwhelm legislators who are attempting to debate laws to restrict trusts. Some of the senators look to the giants for support or advice, while others look away. Perhaps the latter group is as disgusted as the readers of this image who may have been repulsed by the monstrous Mr. Steel Beam Trust and Mr. Copper Trust. Dressed in tails and top hats, the trusts exemplify overconsumption and accumulation while illustrating that monstrosity signified a threat to democracy. That obesity visually represented power would have come as no surprise to Americans who had just experienced the end of Cleveland's first presidency. Until 1908, Cleveland was far and away the largest of America's presidents. Yet, contrary to twenty-first-century norms, Cleveland's weight was not a disadvantage; like the massive trusts, Cleveland's weight signified *gravitas*.

It is not surprising at all that these behemoths should be parodied; after all, these are the very same men—the Astors, Mellons, Rockefellers, Dukes, Vanderbilts, Hills, Carnegies, and Roosevelts—who so unceremoniously threw the masses out of their theaters and stole Shakespeare from the work-

Figure 2.1: Joseph Keppler Sr., "The Bosses of the Senate," *Puck*, January 23, 1889.

ing classes. In the essay "William Shakespeare and the American People," Levine discovered that during much of the nineteenth century, Shakespearean performance was not watched by the wealthy alone. Rather, Shakespeare was popular among all classes of Americans. Toward the end of that century, however, elites appropriated Shakespeare as their own, barring audiences composed of workingmen and women from newly exclusive theaters. Yet, Shakespeare lived on in the popular imagination, if only through the parodies of minstrel shows. Those parodies, notably, did not disparage Shakespeare but instead attacked the pretentious portrayals of Shakespeare by elites. It is at least a little ironic that it was these same elites who simultaneously appropriated Shakespeare from the people and funded modern universities like Carnegie-Mellon, Duke, Vanderbilt, and the University of Chicago that would propel plain speaking over Shakespearian oratory. Before the turn of the twentieth century, however, the political representatives of these elite Leviathans employed the orotund style, speaking very much like the actors that the gilded trust-owners sponsored on the elite stage.

In his 1880 public-speaking textbook, John Poole Sandlands explains the orotund style: "every letter as well as every syllable in a word should be distinctly heard, unless there be some special reason against it."[33] Similarly, in

1882, the actor Dion Boucicault defined orotund speech, "The secret of being heard is not a loud voice.... Every syllable of every word is pronounced, and as far as I can every consonant and every vowel is pronounced.... Now it is the vowel which gives support, and value, and volume to the consonants."[34] Sandlands and Boucicault teach audiences to articulate words by pronouncing every consonant while elongating and quavering vowels to give volume and effect. Before the turn of the century, actors and orators relished consonants, the emphasis suggesting that these sounds were next to divine. Speakers rolled *r*'s, for example, often letting tongues vibrate off the roof of their mouths for a half second. Each letter in the word *our*, for example, was pronounced: ow*wer*, with an emphatically rolled *r*.[35] And pronunciation often mirrored the letters themselves. The *y* in archery, for example, was made to sound like the long *i* rather than *ee*, thus it came out a*rcherrye*.[36] Furthermore, the pace of speech was slow with each word framed by a moment of silence. This style employed constant declamation, sometimes with even the most banal phrase receiving the treatment of a tonally ascending and triumphant-sounding finish.[37] This style may have been necessary to nineteenth-century performance and oration. Shakespearian actors, for example, usually worked in theaters and tents with imperfect acoustics where sound faded before it could reach the farthest listeners. Politicians faced far worse acoustical dilemmas than actors; they often spoke outdoors. Without careful pronunciation, a booming voice, and gestural theatrics that speech teachers taught, how else was one to persuade large audiences?

Election recordings from the 1892 and 1896 campaigns illustrate the orotund style. Candidates did not record their own speeches, letting actors imitate their voices for cylinders that presumably made their way to the nickel-slot phonograph machines found in many saloons. A recording of a Cleveland speech from 1892, for example, sounds ridiculous to twenty-first-century listeners. The voice that leaps out of this cylinder is so over the top that one expects the speaker to experience a heart attack from the exertion. Declaiming on an up note after each clause of the question, "A*re* we↑, their descendants↑, when we have grown to seventy million↑, decla*re* that we↑, a*re* less independent than our forefathers↑?" the speaker answers with a near imitation of a cow "no::↓" that descends into hyperbolic indignity. Sadly, the recording is probably not of Cleveland but is more likely an in-studio reenactment complete with a marching band and an audience, effects that would have been impossible for sound technology in 1892 to pick up at a public event.[38] The actor who imitates Cleveland reiterates what thespians and public-speaking teachers demanded: orotund oratory.

Similarly, an actor mimics an 1896 McKinley stump speech in a recording labeled "From the Front Porch in Canton, Ohio."[39] Like Cleveland, the actor imitating McKinley rolls *r*'s, elongates vowels, emphasizes consonants, and often raises his pitch at the ends of phrases.[40] The sound of applause informs the modern listener that this could not have been the real McKinley speaking from his front porch, though contemporary audiences may have been fooled. As is later discussed about recordings of other politicians in the late 1890s, when Cleveland and McKinley orated, this was how they sounded. That presidential mimics were recorded in 1892 and 1896 tells how little an impression phonograph technology had made on American political culture. That recordings were made at all indicates a growing demand to hear presidential oratory.

Judging by recordings of presidents and vice presidents, the orotund style still prevailed in the last decade of the nineteenth century among the most powerful Americans. But challenges were coming. Henry Irving, a British thespian equally popular on both sides of the Atlantic Ocean, was lambasted by guardians of the orotund style for his attempts to employ vernacular inflections. Critiques of Irving suggest that he did not prolong the sound of his vowels and sometimes employed a common way of pronouncing a word. One critic complained about Irving "vulgarizing poetic tragedy, which should occupy ground removed from the trivialities and the homeliness of ordinary life."[41] Another vented, "Then came a style for which Henry Irving must bear some responsibility, a style in which, in order to escape from the smooth monotony of verse rhythm, the actor deliberately mutilated and wrecked the rhythm, and tried to turn the verse into prose."[42] In the late nineteenth century, oratory was not supposed to mimic the trivialities, monotony, and homeliness of ordinary life. But this was changing as those who lived their lives in prose began to exert political power in a system once dominated by poets.

Until the late nineteenth century, Shakespearian theatrics and political oratory (which had emerged out of Britain's prestige accent) separated speaker from audience and imposed the sound of authority.[43] That commanding sound had formerly substituted for manliness. However, as their power increased, the working class began to challenge the authoritativeness of the prestige accent. At the same time, American men who were engaged in sedentary occupations questioned their own virility and the manliness of elected representatives who spoke in the orotund style. This loss of respect for orotund oratory is illustrated by the feminizing parody of that style in late nineteenth-century vaudeville acts. These shows usually included a bombastic actor, often in the role of "Dandy Jim," who attempted to imitate the orotund style but only managed to botch the message while weakening the authority of the "prissies" who "affected" speaking like pretentious orators.[44] Early twentieth-century recordings made these lampoons, as well as presidential oratory, permanently and universally available. Thus, while voters heard candidates for public office at the saloon and at home (the middle class increasingly being able to afford phonograph players), they also heard those politicians mocked as effeminate and as having voices different from their own. Those voices were no longer manly and thus no longer representative of American manliness.

Present-day viewers might find the appearance, like their sound, of Cleveland and McKinley to be antithetical to an appropriate presidential look. No candidate carrying that much weight could be elected in the current political environment as they would be perceived as lacking in fitness. But it was not a hindrance for Cleveland, who may have benefited from his rotundity. In 1892, when Cleveland campaigned to win back the presidency, cartoonists depicted him as corpulent. Bernhard Gillam's cartoon, "Patching Up the Old Ex-Champion" (1892) is not complimentary yet manages to depict Cleveland participating in the manly activity of boxing (see fig. 2.2). Though Cleveland has been knocked down by the svelte Benjamin Harrison, he is not yet out. Eyes still open, his managers plan a burst of electricity to get him to his feet. Gillam is worried here that though his man Harrison was the sitting president, Cleveland's ability to acquire new charges of electricity meant that Harrison might loose, and indeed that is what happened. The relationship between Cleveland and modernity in the images is striking. Cleveland is plugged into a recent and powerful nineteenth-century invention: electricity. That energy helps to make Cleveland a potent force. Likewise, Joseph Keppler Jr.'s presidential campaign cartoon "Another Explosion at Hand" in *Puck*, September 19, 1900, depicts America's twenty-fifth president in a manly activity. Though he wishes to denigrate McKinley as the toy balloon of William Jennings Bryan, Keppler simultaneously succeeds in trumpeting the authority and manliness of the Ohio Republican. McKinley is dressed in a marshal uniform and, though pompous, carries the *gravitas* of authority as well as his own corpulent body. Given the images of trusts, Cleveland, and McKinley, rotundity could be lampooned, but it still represented authority at least as late as 1900.

The relationship between the visual and the sonic is illustrated by the metaphor of the hot-air balloon in Keppler's image. Hot-air balloons had been employed in America since the Civil War, so the comparison is not simply based on a trendy trope. Instead, it is based on a visual metaphor turned to-

Figure 2.2: Bernhard Gillam, "Patching Up the Old Ex-Champion," *Judge*, September 10, 1892, 172–73.

ward a sonic purpose and then reimagined visually by Keppler. Mark Twain had employed the hot-air metaphor to lampoon the excessive fatuousness of Congress in 1873, a comparison that became very popular starting in 1900 after which the metaphor is expressly vocal.[45] It is the hot-air balloon that gives the actor-turned-emperor his reputation for having (fake) magical powers in L. Frank Baum's 1900 novel *The Wizard of Oz*.[46] The original visual application, real hot-air balloons, became a sonic metaphor for frauds who spoke pompously and became again a visual metaphor for a president who was a little too masculine. The metaphor works so well because it is both sonic and visual and reiterates the connection between the orotund and the rotund.

The Transitional Style

The transition away from the orotund to the common style occurred over decades. Mark Twain, for example, hesitated to abandon official-language conventions. Early handwritten versions of *Adventures of Huckleberry Finn* (1884) indicate Twain's gradual rejection of official language in favor of the common style.⁴⁷ What is most remarkable about *Huck Finn* is that the book was widely popular, as if readers had been waiting to *hear* from characters who spoke normally rather than like the Queen of England. In consuming *Huck Finn,* readers could have imagined the voices of Huck, Jim, the duke, and the king as like their own. However, the duke and the king were con men, who twice put on airs to accommodate their greed. First, they attempted to profit from reenacting a mix of Shakespeare's plays but failed because rather than parody official talk in a vaudevillian skit, they mimicked the orotund style as serious theater. Later, when the duke and king adopt an English accent to cheat a family out of its inheritance, listeners recoil as illustrated by a commonsense town doctor who exclaims, "If these two ain't frauds, I am an idiot." Twain warns the audience through the voice of Doc Robinson not to be taken in by fancy accents, fake actors, and fine oratory.⁴⁸

Perhaps in response, actors began to sing a different tune. No actor better exemplified this transition than Irving, who rejected the overly high tone of his fellow thespians. Critiquing one rival, Irving argued, "What a wonderful actor Wenman would be if he didn't know he'd got a voice."⁴⁹ Against the customs of the theater and the opposition of many critics, Irving achieved a remarkable popularity. Though on the east side of the Atlantic Ocean when Edison first began manufacturing his improved phonograph in 1888, Irving was one of the first to be recorded by the device. A recording of an excerpt of Shakespeare's *Richard III* from 1890 demonstrates the power of Irving's voice and, notably, the ease of his interpretation even more than a century later and on a cylinder riddled with cracks, pops, and fuzz.⁵⁰ Though the words are carefully pronounced, commentators have noted that there is a shift away from an even more stilted form of orotund oration. Irving pronounced vowels in a plainer speaking pattern than his contemporaries and introduced "vulgar" inflections. He might change, for example, the perfectly pronounced "take the rope from my neck" into "teck the rup frum me nek."⁵¹ British and American theater were forever changed.⁵² Thus, in 1918, the actor Louis Calvert reiterated Irving's critique of high-toned actors: "Certainly there is a great danger of becoming infatuated with our faultless diction, of taking excessive pride in it, and of showing it off to the audience. No good actor does that. He never lets the audience think he is speaking beautifully, only that he is speaking naturally and clearly."⁵³ The late nineteenth-century actor Otis Skinner summed up the shift in his 1924 biography, *Footlights and Spotlights.* Thinking about the difference between acting in the 1880s and the 1920s, Skinner remarked, "Romanticism has retired—realism has taken its place."⁵⁴

Speech teachers illustrated a similar shift in the sound of oratory as some began to tolerate elements of the common style. For example, Brainard Gardner Smith advised orators in 1898 to speak as if before friends, though each syllable must still be properly pronounced: "If you speak to your friend as you ought to speak to your audience, your friend will say that you are stilted. Why? Because you must articulate with care."[55] In 1899, Solomon Henry Clark and Frederic Mason Blanchard trained students to speak using the colloquial style, a style that "naturally gives way to a more impressive manner which we may call the elevated style." The colloquial style did not mean, however, "careless and commonplace speaking, but simple, direct, and dignified conversation."[56] In 1903, Edwin Du Bois Shurter also took a moderate line: "As a general principle, we may say that any method of utterance which calls attention to the speaker's pronunciation or enunciation rather than to the thought his language is intended to convey, is a fault. The two extremes of faulty pronunciation are the careless and provincial on the one hand, and the unusual and precise on the other."[57] Shurter announced a compromise between the dialects of white laborers, immigrants, and African Americans and what he termed "affected" oration. If Russell Hunting lampoons the elite chairman as well as the country bumpkin and German and Irish immigrants through a sound recording, Shurter makes the spoof serious by similarly marginalizing the same speakers. But he also argued that oration could not be entirely conversational. Rather, "we must, therefore, draw the distinction between a conversational manner and conversational articulation."[58] Good articulation, according to Shurter, employed exaggerated enunciation so that listeners distant from the speaker might understand each word.

These compromises were produced by concerns about effeminacy. At the turn of the century, male public-speaking professionals were troubled that women were too greatly influencing oratorical style in the nation's homes and schools. Since the Civil War, women had increasingly come to teach rhetoric, purportedly transforming the formerly manly practice of oratory.[59] In 1900, psychologist G. Stanley Hall worried, "Never in the world have so many cubs been so half-orphaned and left to female guidance in home, school and church."[60] Indeed, many writers in this period saw a "crisis in manhood" prompted in part by the dominance of women in the teaching profession. By the year 1900, 75 percent of elementary schoolteachers were women.[61] One year later, in an essay titled "Enemies inside the Elocution Profession," Edward Amherst Ott complained of a problem specific to oratorical instruction: 90 percent of speech educators were women.[62] Male professionals claimed that successful speakers needed "strong," "vigorous" bodies and argued that students should be taught that real oratory was "stern and dignified."[63] "Strong," "vigorous," "stern," and "dignified" are code words for manly. Ott feared the influence of female teachers who, he believed, were rendering male students effeminate. This feminization of public speaking, then, had to be combated by turning away from the sound affiliated with the style of oration—the orotund style—that women taught. That turn favored the common style and in particular the instructional style of the professor who worked in the male sphere of the university.

Recordings of Benjamin Harrison and McKinley's vice president, Garrett Hobart, in the late 1890s provide a hint of the change that appears in contemporary textbooks. Harrison had his voice recorded between 1895 and 1897, probably the first president to do so. His message to diplomats who had participated in a Pan American Congress is a cross between the orotund and conversational style.[64] Like the orator described by Smith, Harrison sounds formal and stilted. Each letter of his prepared text is carefully sounded: vowels are elongated, and consonants are carefully enunciated. He also declaims many of the sentences in the brief recording, employing triumphant up notes to emphasize banal phrases. Yet, his *r*'s are only lightly trilled, and his declamations occur with less frequency than the election recordings of 1892 and 1896. This is a compromise between the orotund and instructional styles. Hobart made a similarly transitional recording in 1898. In his words of welcome for the opening of the Electrical Exposition in New York City, he sounds stilted.[65] He rolls a good number of his *r*'s, particularly in words like "friend," "managers," "wrought," "profitably," "our," "girdle," and "wonder," but more than half of the *r*'s are not rolled. Words like "part" are pronounced in a regional dialect with the *ar* turned into an *ah* sound (think of John F. Kennedy saying p*ah*t). Again, many of the phrases are declaimed, individual consonants are stressed, and vowels are elongated. However, not every vowel is quavered, and not every sentence ends in an ascending pitch, suggesting that even a creature as wedded to the late nineteenth century as Hobart might change his style if only a little.

Frederick Burr Opper's 1900 series of sketches titled "Willie and His Papa" hint at this transition. Opper's cartoons often featured a child named Willie McKinley who dressed like a Shakespearian king and his baby brother, Teddy Roosevelt, who was usually dressed in Western wear. The sketch "What Are You Crying about Now" is reminiscent of Keppler's "Bosses of the Senate" and "Another Explosion at Hand." Papa is "The Trusts," here represented as monstrously large, and is joined by an almost equally gross Senator Marcus A. Hanna in looming over McKinley and his playmate, diminished figures who had been acting in a make-believe play about an emperor and a ship's captain. Notably, Hanna is dressed as a woman, illustrating that obesity was loosing its *gravitas* and was becoming effeminate. But rather than exaggerate McKinley by blowing him up with air, a stream of water reduces McKinley to almost nothing, just as Dorothy reduces the Wicked Witch of the West by dumping a bucket of water on her in *The Wizard of Oz*. Though the elites who had appropriated Shakespeare remained in power behind the scenes and offstage, cartoonists were beginning to promote a more colloquial tone than that of the elite theater. McKinley and his seconds were not keeping up with the times and remained stuck in their Shakespearian outfits.

The Instructional Style

Not surprising, contemporary English playwrights were beginning to compete with Shakespeare for popular acclaim. One of these became a supporter of Sir Irving's modern vocalizations in the late 1890s.[66] Perhaps learning from the elitist attacks that Irving had endured, George Bernard Shaw turned to the theme of dialects in his play *Pygmalion* (1913). In the play, Henry Higgins employs devices, including a phonograph and speech classes, to retrain the working-class Eliza Doolittle so that she can pass in polite society as if she had been born to an elite family. Though Doolittle is able to learn proper speech to perfection, ultimately she gives it up, complaining to Higgins, "I sold flowers. I didn't sell myself. Now you've made a lady of me I'm not fit to sell anything else." Doolittle has discovered that in performing the elite dialect, she has prostituted herself, a less-honest job than selling flowers and one much less humane: "Why did you take my independence from me? Why did I give it up? I'm a slave now, for all my fine clothes."[67] Fortunately, Doolittle has a choice: she can remain a member of polite society by marrying Higgins, or she can marry a more common type, allowing her to revert to her earlier dialect when she wills. She rejects Higgins. The play would have made sense to non-elite listeners and readers. For them, Eliza Doolittle simply stayed true to her roots rather than marry a man who would only be satisfied if she took on airs.

Theodore Roosevelt sounds like the first vice president and first president in decades who did not take on airs. Though he occupied these positions from 1901 to 1909, extant recordings of his voice only date from the election of 1912. In addition to a series of recorded policy pronouncements during that campaign, Roosevelt recorded a mix of two related orations: a stump speech titled "The Right of the People to Rule" and his Progressive Party convention speech, "Confession of Faith."[68] Roosevelt edited the two speeches into a four-minute recording that was intended to reach an audience much wider than the thousands he had spoken to or would speak to in person. Thus, his vocal inflections are designed to persuade listeners as if he spoke before them in a saloon or at home. From the opening, Roosevelt signals that he will not be speaking in the orotund style but rather like those Americans who would soon be voting. During the recording, Roosevelt quavers not the vowels but instead accentuates consonant-vowel pairs.[69] For example, when he informs his audience that his opponents betray their beliefs through the "*way* in which they champion every device to make the nominal *ru*le of the people a sham," he stresses phonemes that earlier orators would not have (think of Walter Cronkite saying, "And that's the way it is"). Without rolling his *r*'s—in four minutes, only two are trilled—he equalizes vowels and consonants rather than elongating the former while emphasizing the latter. Roosevelt also fails to pronounce every consonant and syllable. The word "pleasure," for example, becomes "plezha," just as "in order" becomes "inahda." Late nineteenth-century speech teachers had urged orators "to give the consonants their proper value, particularly when they stand at the end of words."[70] Roosevelt indulges in violating this rule. A similar drop-off occurs when Roosevelt describes the problems of "unreasonable conservatism and unreasonable radical °ism°," the last, "ism," being nearly inaudible.[71] Roosevelt also fails to pronounce each word distinctly, occasionally eliding sounds that intruded on the rhythm of the speech as when he drops the *d* in "blind reaction," turning the phrase into "blinereaction." Rather than orate like Shakespeare, Roosevelt talks like the people.

Like a professor of history, Roosevelt recounts a violent episode from the French Revolution.[72] The bloodshed resulted because "the beneficiaries of privilege, the forward reactionaries, the shortsighted ultraconservatives turned down↓ Turgot:". At the phrase "turn down↓," his pitch falls, making the words a kind of reverse onomatopoeia by which the intonation sonically creates meaning that reiterates the word. Here, Roosevelt also employs a storyteller's trick; he holds the audience's attention by continuing the last vowel of the sentence (Torgoe:) almost until he begins the next phrase, denying auditors a natural break where they might stop listening. Indeed, he sounds like a storyteller who spins tales by the light of a campfire. That Roosevelt must have been a storyteller is apparent both from his anecdote-filled books and his use of frontier mythology in his political rhetoric.[73] In *Ranch Life and the Hunting Trail*, for example, he employed dime-novel storytelling conventions in describing his experience as a Western gunslinger.⁷⁴ Roosevelt understood that storytelling connected with listeners, and that meant plain speaking.

Roosevelt had yet another sonic technique for capturing his audience's attention: musical metaphor. Abandoning the headiness of the orotund habit of ascending the pitch at the end of phrases and sentences, Roosevelt's pitch sometimes musically descends. Given that Roosevelt recorded the 1912 speech after public-speaking teachers had paired speech with music, one should not be surprised to hear the former president using the piano as a sonic metaphor. In 1890, Theodore Emanuel Schmauk taught, "Melody pleases and stirs by an awakening and responsive power in itself, while words affect the understanding chiefly by the power of a conventional signification which has been placed in them. . . . The twin arts, however, cannot afford to be independent of each other. The best song needs a basis of thought; and the best speech needs a music of utterance."⁷⁵ Similarly, Ernest Pertwee wrote in 1902, "The study of singing is a considerable aid to the master of the speaking voice. The vocalizing of scales, up and down the compass, using the open vowels AH, OH, OO, and after a time gradating and shading the tone diminuendo to fortissimo and back to diminuendo . . . is an excellent discipline for 'forming' the speaking voice."⁷⁶ And David Ffrangcon-Davies rejected the separation of "the phonetics of the singing voice from those of speech" in 1905.⁷⁷ For turn-of-the-century speech teachers, to speak well was to employ musical figures. Thus, toward the beginning of "The Right of the People to Rule," Roosevelt drops his tone through three successive clauses. Though he does not sing, Roosevelt chooses the key of C to argue that the American people are fit "to govern themselves [pitched in the scientific designation at G_4], to rule themselves [E_4], [and] to control themselves [C_4]." Later in the recording, he repeats this sonic device when he argues, "Our aim must be steady [B_5], wise [G_5] progress [E_4]."⁷⁸

Notably, visual corollaries attest to Roosevelt's musical talent. The cartoon "Pianissimo Teddy" features the president at the piano, accompanied by Columbia (who represents America) and amid portraits that depict a middle-class parlor rather than a highbrow concert hall (see fig. 2.3). Though in a bow tie and tails, Teddy plays from *Fortissimo Strenuouso,* a fictional "musical adaptation" of his best-selling book *The Strenuous Life.* Likewise, another cartoon, "Progressive Fallacies," depicts Roosevelt singing "better than Bob [Taft]" in yet another parlor, this one equipped with a sofa.⁷⁹ Perhaps the cartoonists knew that like millions of American families, the Roosevelts turned to the piano and singing for relaxation and solidarity. At the beginning of the twentieth century, the piano was a common feature

Figure 2.3: McKee Barclay, "Pianissimo Teddy!" (Baltimore) *Sun*; repr. in *T. R. in Cartoon*, ed. Raymond Gros (New York: Saalfield, 1910), 28.

of working-class saloons and middle-class parlors across America.[80] Many parents, including Roosevelt, were accustomed to hearing children learning their scales on the piano keys. And most would also have heard popular tunes on a daily basis, many of which were written in the key of C. Though he may not have known about the ancient relationship between music and rhetoric, Roosevelt vocalizes in the key of C, and thus he fits expectations about pitch. His voice would have been as common as the sound of a child learning to play piano. He had figured out how to sound like friends at the bar and family at home.

He was probably compelled to adopt this style to overcome his past. Much of Roosevelt's public persona was crafted in reaction to accusations of effeminacy lodged against him by New York City's press in the 1880s. Newspapermen had accused Roosevelt of being a weakling, a Jane Dandy, a Punkin' Lily, "the exquisite Mr. Roosevelt," "Oscar Wilde," and even a fellow who was "given to sucking the knob of an ivory cane."[81] Fearing harm to his reputation and political career, Roosevelt overcompensated by becoming a big-game hunter, Wild West sheriff, and colonel of the Rough Riders. A contributing factor to the early allegations may have been Roosevelt's high and soft voice. Thus, Roosevelt modified his speaking patterns in a manner that the public would accept as appropriately manly.[82] In this effort, he received support from speech scholars. In *Public Speaking and Reading* (1896), Edward Napoleon Kirby informs readers that men with high-pitched voices could maintain their authority by sounding intellective and didactic.[83] It is exactly this style that Roosevelt employed in the 1912 recording, a style manifested in a cartoon that depicts Roosevelt teaching Uncle Sam proper pronunciation: "KIST, not KISSED" (see fig. 2.4). Moving away from the orotund convention of pronouncing even the "ed" at the end of words, Roosevelt teaches his pupil to elide the ends of words and to sound like the people.

In campaign recordings from 1908 and 1912, Roosevelt, Bryan, Taft, and Wilson all speak in a style that one might today associate with instruction.[84] Scholars who have studied formal addresses have argued that politicians have employed a conversational rhetoric since the beginning of the twentieth century. As the foregoing indicates, the recorded orations from 1908 and 1912 are better described as instructional.[85] Rather than engage in a dialogue with the people, these presidents sound like they perceived themselves as teachers. And though twenty-first-century students no longer associate manliness with the professoriate, Yale and Harvard men at the turn of the twentieth century, with their experience on football teams, their successful careers, and their traditions, were perceived as bastions of heroic manliness.[86] Through the content of his speeches and in particular the intonation of his oratory, Roosevelt sounded the new instructional, manly style.

Tacked onto the McKinley presidential ticket of 1900, Roosevelt brought energy and virility to the campaign. In many ways, that ticket represented the old and the new, the orotund oratory of the nineteenth century and the instructional style of the twentieth century. After Teddy Roosevelt, Americans expected their presidents to be physically fit, and those who weren't would struggle with press coverage of their diets and exercise regimens. Not

Figure 2.4: J. L. De Mar, "Roosevelt as a Schoolmaster" *Philadelphia Record,* 1906; repr. in *T. R. in Cartoon,* ed. Raymond Gros (New York: Saalfield, 1910), 212.

surprising, the imagery of fat politicians became increasingly effeminate after Roosevelt's succession to the presidency in 1901. William A. Rogers in his cartoon "Congress on His Hands," for example, complimented Roosevelt's strenuous regimen while lampooning the mass of Congress (see fig. 2.5). As that body is incapable of moving legislation to reform Cuba, finances, labor practices, and the trusts, Roosevelt carries their measures as if he is a bridegroom stepping over the threshold. Similarly, in the cartoon "Home Again" (1905), Rogers illustrates Roosevelt's return to Washington, D.C., from a strenuous vacation of hunting cougars, wolves, and bears and of roughing it with only a rifle and sleeping mat.[87] In his absence, the Capitol has become a boiling pot (a visual cue for the kitchen and thus the woman's

Figure 2.5: William A. Rogers, "Congress on His Hands," *Harper's Weekly*, November 14, 1903.

sphere) of scandals. While Congress had hoped to keep the lid on things by placing a rotund monster on top, Roosevelt has discovered the mess and prepares, hands on hips, to clean up. Shot upwards, the monster is powerless in the face of a man who is firmly grounded and wielding a firearm.

Fitness

Roosevelt's successor spent his formative years in a culture that valued the physical manifestation of *gravitas* but became president at a time when attitudes toward obesity had changed. In 1908, he no longer fits. In Rogers's "You'd Look So Much Better in Your Own Clothes" in the June 13, 1908, *Harper's Weekly,* Taft is compared to the absent Roosevelt through an ill-suited military uniform. As Taft struggles to maintain his composure in his too-tight outfit, a fit and thin Uncle Sam looks on, laughing at the scene while signaling the audience that they, too, are permitted to laugh at their future president. At every extremity, Taft's clothes stretch to impossible limits, suggesting that the man underneath is unfit for the presidency. And where earlier cartoonists had drawn attention to Cleveland's and McKinley's manliness by emphasizing the crotch, Rogers makes Taft's crotch disappear under a massive gut and the unseen posterior that pulls all the weight in the lower extremities toward a place beyond the reader's ability to see. Upon the installation of Taft as president, some cartoonists reflected a belief that the new leader would not be as manly as the retiring president. In Samuel D. Erhart's "Baby Kiss Papa Good-bye" in the February 24, 1909, *Puck,* Taft and Roosevelt stand in contrast (see fig. 2.6). A fit Roosevelt, attired in a hunting outfit, is preparing to leave for new adventures. A servant carries his big stick. Indeed, the only phallic symbol in Washington, D.C., that was larger than Roosevelt's was the 555-foot-tall monument (completed in 1884) that testifies to the manliness and fitness of another military leader and president, George Washington. Taft, on the other hand, is no Washington or even Roosevelt. Rather than being dressed for the hunt and carrying a big stick, Taft wears a frilly maid's apron and hat and carries Roosevelt's baby, My Policies. Unfortunately for Taft, new ideas about presidential fitness left him possessing weight but no *gravitas.*

"Goodness Gracious! I Must Have Been Dozing!" in the June 1910 edition of *Puck* completes the feminization of Taft. The twenty-seventh president is still dressed in the domestic uniform of the maid, though his apron and hat have lost their frills as well as their crispness.[88] Taft has been in power for a year and a half now and has been sleeping on the job. His sewing kit, designed to imitate the Capitol, is tangled in a mess of yarn that has been

Figure 2.6: Samuel D. Erhart, "Baby Kiss Papa Good-bye," *Puck,* February 24, 1909, cover.

mischievously unspun. Four frolicking pussycats have produced the mischief. Overlooking the tangled web is none other than the former president himself, reminding us again of the difference between Taft's femininity and Roosevelt's manliness. By 1910, fat visually marked a lack of fitness for the presidency.

The argument I make here about the image of the presidency, however, is subservient to a larger argument: that sound can be read. What does reading sound get us? In reading the sound of presidents around the turn of the twentieth century, we discover supporting evidence for other historical events. Evidence that expectations about the physical appearance of the presidency changed around 1901, for example, is buttressed by sonic evidence of a change in the sound of the presidency. We have both seen and heard that this change was prompted by the desire of elites, exemplified by Roosevelt, to be perceived as manly by middle- and working-class voters, who became increasingly aware of how presidents looked and sounded during this period of American history.

Reading the sound of presidents provides support for an hypothesis that rhetorical scholars, historians, and political scientists have put forward about the relationship between the president and the people at the turn of the twentieth century. These scholars have argued that beginning with Theodore Roosevelt, presidential rhetoric takes a turn toward the popular, culminating with the election of 1912.[89] These scholars do not examine the sound of the presidency, focusing instead on the words employed in presidential address. They argue that these orations verbally become more democratic and colloquial after 1901, an argument that is buttressed by the sound recordings of presidential oration on both sides of the turn of the century. Not only did presidents and presidential candidates shift toward the use of common words and phrases but they also shifted to a common style of speaking. The turn toward a less-elite form of oration reminds us that the power of the working class and middle class had grown by the turn of the twentieth century and in conjunction with new technologies like the phonograph enabled the consumption of political rhetoric in a manner that few Americans had previously experienced. If presidential candidates were to be heard through phonographs in saloons and parlors across America, then they had better sound like they were in the saloon or the parlor and not in the highbrow theater.[90]

In reading the sound of the presidency from Grover Cleveland to William Howard Taft, it becomes clear that cultural assumptions concerning manliness and authority can be heard in the sound of presidential oratory. Reading the sound of such speeches allows us to hear the influence of the common style on the presidency and in turn how during the recorded age the sound of the presidency shifted in a manner that represented idealized American manhood. No figure better illustrates that shift than Abraham Lincoln, whose Gettysburg Address was performed by Len Spencer in a 1903 recording.[91] Invoking the sound of a Shakespearian performance, Spencer hammed up every syllable and savored every vowel and consonant. But Lincoln's original popularity was at least partly predicated on a homespun manner that reflected his Kentucky, Indiana, and Illinois roots. What is particularly damning to any claim of accuracy for Spencer's rendition is the actor's resort to a deep voice. Spencer and his audiences believed their greatest president to have expressed himself in a clear baritone, thus reflecting American authority and manliness. This late nineteenth-century myth was intended to turn Lincoln into an appropriately manly president. Yet, first-hand accounts indicate that this president spoke in a shrill voice.[92] Spencer's deep-voiced performance does not overcompensate for Lincoln's shrillness, rather it overcompensates for the orotund style that was rapidly losing its authority. In 1898, William Pettenger had railed against the common style in the form of "the sing-song voice, the lisp, the guttural and tremulous tones, the rhythmical emphasis which falls like a trip-hammer at measured intervals [as] specimens of common, bad habits that should be weeded out as fast as they can push through the soil."[93] Ultimately, however, it was orotund speaking that came to be associated with the sing-song voice and the lisp. After 1903, that style was weeded out as fast as it could push through the soil.

In the first decades of the twentieth century, the orotund style disappeared from serious recordings. In 1903, listeners may still have believed that Lincoln spoke like a late nineteenth-century thespian, but by 1919, sonic expectations had entirely changed. Spencer's Shakespearian take on Lincoln thus gave way to a Lincoln who spoke like a commoner. Training students to orate by reciting the Gettysburg Address, Frederick Houk Law lectured students to slur the words "and," "upon," "a," "in," "to," "the," "that," and "are."[94] Where Spencer had declaimed the Gettysburg Address syllable by precious syllable, Law encouraged students to take a more colloquial approach if only to emphasize important words like "nation," "liberty," and "equal." By D. W. Griffith's talkie *Abraham Lincoln* (1930), the imagined Lincoln could be seen *and* heard employing the common style.[95] Between the two imitations, audiences had acquired new assumptions about the sound of the American presidency. One *New York Times* reader gushed in 1933 about the vocal style of two American presidents:

> Due recognition is therein given to the simplicity of [Franklin Roosevelt's] language—his "plain, straightforward, colloquial English"—his "words that all could understand"—his "voice that reveals sincerity, goodwill and kindness, determination, conviction, strength, courage and abounding happiness." Thus it was, also, close upon seventy years ago, with a man who stood upon the Gettysburg battlefield. Abraham Lincoln there addressed his fellow citizens. He had no magic radio by which to spread his inspired words abroad, but he personally spoke with all those tremendously important qualifications that THE TIMES writer today hails in his successor as President.[96]

By 1933, Franklin D. Roosevelt had mastered the common style demanded by mass audiences. Apparently, so, too, had Lincoln. He was no longer the orotund thespian of 1903 but had become a speaker who expressed himself in "plain, straightforward, colloquial English" and in so doing revealed through his voice that he was a man of sincerity, goodwill, kindness, determination, conviction, strength, courage, and abounding happiness. That Lincoln might reveal his abounding happiness on the Gettysburg battlefield reminds us that assumptions about the presidential sound can be deafening.

The study of sound is central to understanding presidential oratory and rhetoric more broadly. This kind of analysis provides evidence and nuance to arguments that scholars have made about the ways presidential rhetoric serves to define (and is defined by) cultural expectations of American-ness. In this instance, pressures to shore up manliness can be heard in the sound of the presidency at the turn of the twentieth century. If we can learn to hear sonic expectations of another age, perhaps we can learn to hear our own as well. Today, for example, Americans are still likely to vote for presidents who express sonic manliness as it is currently defined. Listening to the sound of dialects and vocalizations furthers our understanding of the persuasive power of political speech and sonic media while complementing published scholarship about words and images. In doing so, we add a microphone to our window on the past.

3

Machine Mouth

At the end of the nineteenth and beginning of the twentieth centuries, many found the sounds of modernity disturbing. But humans are remarkably adaptable, and so over time and sometimes with training, most grew familiar with the sound of the ticking clock, the clickity-clack of the locomotive, and the ceaseless humming of countless machines. Did this shift occur without remark? Or did people write about being disturbed by the sounds of modernity? As it turns out, much has been written about the sounds of modernity, both before and after people adjusted to the new noises. Three inventions in particular were often remarked upon by those who survived the transition into modernity: the clock, the factory machine, and the locomotive. Though intended to be precise, the sound of the clock, for example, produced imprecision and madness as sufferers responded to tick-tocking, tolling bells, and ringing alarms by attacking clocks, people, and other instruments associated with time.

Thanks in large part to the railroad interests, the clock and precise time became an unstoppable force in America and elsewhere, prompting violent reactions against its noises. Literary creatures like Richard Wright's Bigger Thomas, from the novel *Native Son,* rebelled against clock sounds, illustrating that even in 1940, not every American had come to grips with industrial time and its noises. The shattered soul of Bigger had precursors in the visual and sonic art that appeared before World War I. Cubists, Futurists, and hypermodern composers like Igor Stravinsky celebrated industrial noise by employing sonic motifs to penetrate and shatter the self into thousands of shards, turning humans into barely recognizable patterns. These attempts to perform the fragmentation of modern selves met with resistance after the horrors of World War I, a resistance epitomized by Fritz Lang and Thea von Harbou in their 1927 film *Metropolis,* in particular during a scene in which modern workers march willfully into the gaping mouth of a machine. Though the visual and sonic arts witnessed a return to order in the years after World War I, migrants like Bigger who fled preindustrial rural areas remained behind the times, not yet accustomed to the sounds of modernity and still prone, even in 1940, to being shattered by mechanical noises.

But sound could also protect and unify. Adolf Hitler, not surprising, represents both the potential of noise to shatter selves and simultaneously the power of sonic technologies to protect selves. His deft manipulation of microphones, radios, film, and crowd noises turned the German people into an aggressive and unified, if blind, force. The bluesman Bukka White presents a better alternative to Wright's violent novel and Hitler's manipulative voice. Published in 1940, White's song "Special Streamline" employs a storytelling style and guitar melodies to teach listeners how to adapt to modern, industrial tempo and noise. By imitating locomotive sounds and by informing his audience that those motifs can be turned into music, White helped his listeners adjust. The clock and the train share an intricately connected history. Both have been described as synecdoches of modern industry, and both have helped the other to dominate the American soundscape. Once, these disturbances produced madness and violence. By the 1950s, they had become iconic and comforting sounds.

Wright and White are the most apt interpreters of sound that shatters and protects. They were both from Mississippi (Wright having migrated north), both were fans of boxing (White was himself a boxer), and both were bluesmen. Though Wright was not a songwriter, his literary works were called blues by Ralph Ellison.[1] Wright even penned blues lyrics for a musical tribute to world champion boxer Joe Louis that was performed by Count Basie and Paul Robeson in 1941. The song "King Joe" exemplifies the experience of industrialization:

> Old Joe wrestled Ford engines,
> Lord, it was a shame;
> Say old Joe wrestled Ford engines, Lord, it was a shame;
> And he turned engine himself and went to the fighting game.[2]

For Wright and White, the key question of 1940 was whether southern blacks could adapt to modern, industrial time or become machines themselves.[3] Wright's protagonist in *Native Son* is so disturbed by industrial modernity that he responds by murdering two women. Similarly, White spent part of his life behind bars in the euphemistically named Parchman Farm because he *had* murdered a man.[4] But Wright was not Bigger, and White was not Bukka (in fact, he was named after Booker T. Washington). Both emerged from the violence as adapters of industrialization: Wright through the novel and White through the song, as each explored the modulation of industrial noise into blues. By modulating industrial noise into the verbal and sonic lexicon of the blues, African Americans learned to adapt to modernity.

This chapter builds on a lesson learned in the previous chapter: sound can be read. A president's voice carries more rhetorical force than the words on a page we are accustomed to studying. Can this lesson, then, be applied to sounds that are not voices, the sounds of machines?

Shattering Noises

As Lewis Mumford noted in 1934, "the clock, not the steam-engine, is the key-machine of the modern industrial age. For every phase of its deployment the clock is both the outstanding fact and the typical symbol of the machine."[5] According to Mumford, the clock was the synecdoche of modernity. In previous ages, the church bell had imperfectly observed time, set for the most part by watchers of the sun's zenith as it marked noon on local sundials. Though imprecise, those bells, according to a variety of sound scholars, operated to create unique and comforting sonic environments, "soundscapes," in parishes across Europe and the Americas.[6] The introduction of mechanical clocks made the bells sound at slightly more precise hours even as the clocks continued to run fast or slow. By the eighteenth century, costly clocks and watches had become fairly accurate. That improvement, however, did not begin to reach most Americans until after the manufacture of inexpensive yet nearly accurate watches started in Connecticut in the 1850s.[7] As the watch and family clock became increasingly common, other forms of time also penetrated into the daily lives of Americans. For example, industries rapidly introduced the punch clock after its invention in 1888 by a company that would eventually become IBM. Those clocks disciplined workers into arriving and leaving "on time," a term that transformed time into a physical and geographical reality.[8] As America's railroad network spread, the bells, whistles, and steam explosions of the on-time train also began to signal precise hours in communities both small and large.

Led by the railroad interests, industrialists embraced accurate, mechanical time. For railroad corporations, precise time was necessary for survival. Without the imposition of standard time zones in 1875, for example, trains would have continued to collide because time at one departure did not match time at another. The new discrepancy between standard and local times produced by this innovation gave one future industrialist his start: a young Henry Ford made double clocks that could be set to provide both times. For all industrialists, engines and workers had to be made timely. Thus, public-school students learned about time, and workers were disciplined to its regimen.[9] In the 1880s, Frederick Winslow Taylor promoted a management style, "Taylorism," that Ford and others adopted for their factories. The "scientific" management of time and labor dictated the adoption of strict time in all facets of the workplace. As early as 1847, Karl Marx had anticipated the scientific management of time, man, and machine: "Through the subordination of man to the machine the situation arises in which men are effaced by their labor; in which the pendulum of the clock has become as accurate a measure of the relative productivity of two workers as it is of the speed of two locomotives."[10] For Marx, man is effaced as clocks increasingly control and measure locomotives and workers, treating both as engines that served a new, modern commodity: time. Not surprising, by the early twentieth century, industrialists were comparing their laborers to clocks.[11] This imposition of time helped to produce and entrench modern industries and bureaucracies. As Walter Benjamin and Benedict Anderson argue, homogenous empty time, an imagined universal time, is critical to narratives of progress and the modern nation-state.[12]

Notably, clocks disturbed not because of their appearance but rather because of their sound. One could always look away, but as sound theorists know, one can never close one's ears.[13] Against this background, many challenged time. Folk traditions maintained a literal timelessness that resisted the precision of mechanical time. African Americans, in particular, bent time with a genre of music that emerged during the interwar period as jazz. Thanks to the mechanical crank that turned recording devices, Thomas Edison and his consumers sped up and slowed down phonograph machines, warping time by distorting voices and music. Even employers manipulated whistles and clocks to begin the workday early and end the workday late.[14] That manipulation has its fictional projection in *Metropolis*.[15] In that film, the protagonist takes the job of one of the dystopian underworld's laborers, spinning the arms of a clocklike machine for an unexplained reason. As he nears the end of his shift, the machine transforms into a clock. The protagonist tries with all of his might to push the arms, set to five minutes before the hour, forward so that he can get some rest. As the musical score emphasizes the pounding click of each second, the clock's arms become stronger than the protagonist's and move counterclockwise, twisting him around. His hands tied to the arms, he is crucified upon the machinery of modernity.

The constant ticking of these newly ubiquitous machines drove some to distraction. By the turn of the twentieth century, clocks incessantly and noisily ticked away, endlessly reminding laborers and the middle class that they were subjects of time. The ticking, the bells, and the alarms produced violent reactions. In 1940, Benjamin quoted approvingly from an 1830 French poem:

> Who would have thought! we say
> Who are irritated by the hour,
> These new Joshuas
> At the feet of each clock tower,
> Shot at the faces
> To stop the day.[16]

Joshua was a leader in the biblical account of Jewish slaves freeing themselves from their Egyptian masters, and so "new Joshuas" is a metaphor for wage slaves who rebel against the tyranny of increasingly modern time. The poem anthropomorphizes the clock towers (feet, faces) as slave drivers who disturb the new Joshuas. The irritation caused "by the hour" is not a visual provocation but rather sonic. The Joshuas revolt against the sound of the striking of each hour and thus against the noise of time. Even as late as 1954, the venerable Sherlock Holmes could sympathize with a man who attacked clocks not by their image but by their sound. The poor fellow did not see the clocks he attacked, rather he swung a club at places where the sound of a clock emerged.[17] These modern Joshuas had not yet learned to internalize the sound of time, and thus that noise remained threatening. Wright's protagonist in *Native Son* exemplifies this violent reaction against the noise of time.

In this novel, Wright's sonic motifs illustrate affective experience. The blues form, argues the novelist, is an affective reaction to a threatening environment.[18] Humans react to sound, rather than act upon sound. Similarly, Lawrence Grossberg argues that sound in and of itself is affect. It has no meaning and cannot be interpreted, thus listeners have no rational agency in their responses to it.[19] Instead, listeners react emotionally and instinctively to sound. Grossberg's argument is derived from Plato's and Kant's belief that music provokes emotion and is irrational. Sounds, argues Grossberg, are preideological and thus can prevent ideologies from forming or can buttress the formation of ideologies. Bigger Thomas exemplifies the clash between the sonic production of affect and reification of ideology. Noises prevent him from adapting to modern Chicago and confine him to a preindustrial state.

As evidenced by the titles of the first two "books" in the novel, he experiences only fright and flight. But Wright, as author, is not deprived of agency by the sounds of clocks. Though communication comprises symbols *and* sensory impressions, these may be managed by actors like Wright who think beyond words.[20] Thus, Wright manages the sounds of the clock, turning these into signifiers of self-shattering while producing a story about adapting to industrial noise.

Wright's novel announces itself with an alarm clock that wakes up the Thomas family. The clock foreshadows a tale about the destruction of three souls because of modern industrial noise and in particular the clock. Bigger's first attempt at a crime, an armed robbery against a white man, is predicated on time: a clock that he sees through an open window, and a window of time when a police officer will be away from the targeted store. But the attempt fails because one of Bigger's accomplices is late. Understanding that success means being on time, Bigger complains, "Every time somebody's late, things go wrong. Look at the big guys. You don't ever hear of them being late do you? Naw! They work like clocks." In modern America, Bigger understands, success of all kinds, even in criminal endeavors, requires working like a clock. After Bigger is forced to take a job with a family of white capitalists, he must confront time with stunning regularity. For example, Bigger worries that his cheap wristwatch is inadequate now that he works for white people, a reference to the inaccuracy of the watches the poor could afford and to the timelessness of the culture that Bigger had to abandon. Thus after committing his first crime, the murder of a young white girl, Bigger fears that modern industrial time would catch up to him because it moved so much faster than his own.[21]

Indeed, the awful crime was committed because of the conflict between Bigger's agricultural sense of time and modern, industrial culture. The novel opens with the onomatopoeic "*Brrrrrrriiiiiiiiiiiiiiiiiiinng!*" Elsewhere, "a clock boomed five times," while another clock disturbed silence with a slow tick. When Bigger commits his first murder, time is of paramount interest. One particular clock sonically informs Bigger of when the murder happened, ticks again to remind him that "time was passing," and reappears at the end of the novel when he is forced to confront his crimes and his execution. The clock stands as his accuser and as a witness to his execution. His own alarm clock, the one that loudly opens the novel, almost betrays him because it tells the difference between the time of the murder and the time of Bigger's false alibi. He returns to the scene of the crime "to be there with his fingers on the pulse of time." He finds out too late that he has failed to read time correctly because the boyfriend of the murder victim and the murder victim herself have prevented the white family's household from running like a clock.[22]

Ancient time in the form of a tolling bell that disturbs his dreams connects with modern time in condemning Bigger: "Out of the surrounding silence and darkness came the quiet ringing of a distant church bell, thin, faint, but clear. It tolled, soft, then loud, then still louder, so loud that he wondered where it was. It sounded suddenly directly above his head and when he looked it was not there but went on tolling and with each passing moment he felt an urgent need to run and hide as though the bell were sounding a warning." Because he has only been schooled through eight grades and has not learned to internalize the sound of the modern clock or the tolling of the bell, the alarms, the chimes, and even the tick-tock leave Bigger affected. He knows that working like a clock is critical to his success, but his efforts are imprecise and untrained. Time fights Bigger. It fails him when his cheap wristwatch stops, predicting his capture by the police and ultimately his execution. Even the city newspaper colludes with modernity. Wright spells the daily out only as the *Times*. Just as the broken wristwatch predicts Bigger's capture, the *Times* informs Bigger of when it will happen. In the end, he succumbs to modern time, signing a confession to his deeds that prompts a jailor to proclaim, "He came through like a clock." Has Bigger finally adapted? A second jailor believes otherwise: "[He's] just a colored boy from Mississippi."[23]

Bigger's sense of time failed to sync with the industrial time that he entered into when he moved to Chicago and began working for a white, capitalist family. The time that he learned in agrarian Mississippi was nothing like that of the industrial north. The tempo of life had changed. But, we might ask, what is so maddening about clocks? These are always present after all, and from the moment school starts to the moment we retire, these dictate the pace at which we live. But this was not so for French workers in the 1830s, and it was still not so for Bigger Thomas in 1940. Clock sounds affected them. The tick tock, the "*Brrrrrrriiiiiiiiiiiiiiiiiinng!*" of the alarm, and the booming of the hours penetrated into their consciousness and shattered their selves. Their reactions may seem two-dimensional, given that Wright, for example, denied his protagonist agency in favor of buffeting Bigger in a tempest of emotions.

Indeed, Wright literally and consciously signals affect in this novel through the use of both the clock and an urban locomotive—the streetcar. The streetcar rattles and rumbles through Bigger's mind accompanied by the hissing of radiators and fire-truck-pumped water.[24] Bigger's ears are also penetrated by the rattling of coal in the chute, the whirring, heaving, and throbbing of automobile engines, and other industrial noises—a veritable symphony of modernity.[25] As Wright notes, "Bigger felt that he was caught up in a vast but delicate machine whose wheels would whir no matter what was pitted against them."[26] The whirring would not stop but would continue to trigger the protagonist's affective responses. Bigger has no agency and succumbs to his emotions.

Importantly, Wright ties locomotives to the clock and other modern, industrial sounds through one particular noise. After the onomatopoeic alarm that starts the novel, Wright signals, "An alarm clock clanged in the dark and silent room." Bigger has been irrevocably disturbed by a clang. When he is asked to "shut that thing off," Wright describes how the "clang ceased immediately." "Clanging" sounds again when Bigger has his nightmare about the church bell tolling the hours to warn him against the future. Though the sound of the clock is a dominant theme in the first part of the novel, gradually the clanging sounds of other industrial noises take primacy in the tale. The clang of the streetcar leaves Bigger alone with his first victim. We hear the clanging of shovel on stove when his crime is almost discovered and later in the form of emergency vehicles at the conclusion of the section titled "Flight," signifying his capture. The clanging of passing streetcars occurs again as Bigger lies to the police. And that clanging reappears as the noise of his prison door repeatedly closing.[27] Indeed, just as the word *clang* appears at the very beginning of *Native Son* (it is the fifth word of the novel) to signify an alarm clock at the beginning of time or perhaps the bell that signals the start of a boxing match, the noise sounds again at the end of the tale (it is the second-to-last word of the novel) to signify a prison door slamming shut on Bigger's life—the end of time. Bigger has been knocked out. From beginning to end, Bigger is affected by modern, industrial clanging.

The use of a single onomatopoeic word *clang* to signify a wide variety of industrial noises—clocks, streetcars, coal shovels, emergency vehicles, and prison doors—all of which could be described with other words, illustrates the importance of industrial noise to the development of both the novel's plot and Wright's adoption of modernity into the blues. Elsewhere, industrial noises are put to other uses. *Clack*, for example, describes the sound of pool balls colliding (35), *clamor* with the sound of crowds, *clatter* with the sound of objects colliding with floors and walls,[28] and *humming* with a musical instrument and a fire. In short, Wright distinguishes between onomatopoeic industrial noises (clang, hiss, throb, whir) that frighten Bigger and other noises that could also be used to describe modernity (like clack, clamor, clatter, and humming) but that are not used for that purpose. Notably, the industrial noises also signify engines that have accelerated the pace of life: clocks, automobiles, streetcars, and so on. The sounds produced by this quickening are not only soul-shattering themselves but are also synecdoches of the accelerating velocity of humanity, a velocity that also threatens to tear souls apart.

Scholars argue that by the modern age, the body had become a barrier against the outside, containing a perfectly formed self that could no longer be penetrated or possessed. Before modernity, people believed that the body was open to outside forces; that it could be penetrated by noise as well as by spirits, a phenomenon most spectacularly apparent in stories of possession and exorcism.[29] That belief reappeared, though less consciously, when the new noises of modernity—ticking clocks, rumbling trains, whirring machines— were disturbing and maddening. In an essay titled "Noises," Charles Dickens complained in 1871, "There is some 'machinery in motion' against my study window.... [A builder has] a driving wheel with an endless strap on the tire. That wheel is driving me to distraction. That strap has entered into my soul. He has set up a circular saw—twenty circular saws I should say. They are sawing my heart in twain."[30] Describing the 1932 Freudian novel *Voyage au bout de la nuit*, the historian Joel Dinerstein explains the effect modern noise has on the protagonist, who "equates the physiological experience of the factory roar to possession by a mechanical god."[31] As the machine age began to dominate the lives of Europeans and Americans, modern noises threatened to penetrate into and possess the self again. Gilles Deleuze and Felix Guattari describe how machinic noises transpierce and deterritorialize us.[32] Sounds enter into and disturb our wholeness. To be penetrated by sound is to lose one's sense of unity and to be unable to differentiate one's self from the outside world and from others.[33]

Illustrating the threat of machinic noise to self-hood, nineteenth-century novelists described how the din of factories replaced and prevented the speech of humans. Herman Melville describes a modern factory: "The human voice was banished from the spot ... [as workers were] feeding the iron animal."[34] For Rebecca Harding Davis, "the clamor begins with fresh, breathless, vigor, the engines sob and shriek."[35] Dickens further describes the teeth of a mechanical saw that created a "shrill, screaming, ceaseless whirr."[36] Similarly, Thomas Hardy describes "the inexorable wheels continuing to spin, and the penetrating hum of the thresher to thrill to the very marrow all who were near the revolving wirecage."[37] Even as late as the 1980s, Andean Indians described engine noises that screamed, sucked, and whined.[38] It is remarkable how consistently the metaphor of the mouth and the voice have been applied to the din made by machines. For the twenty-first-century reader, the *hum* of machinery is mostly background noise, even if loud. It is usually not penetrating. Indeed, the very word *hum* no longer has the negative connotation that authors assigned it in the nineteenth century. For Melville, Davis, Dickens, and Hardy, those noises threatened to dehumanize workers and to rob them of their selves. Anthropomorphized into a loudmouth, Hardy's thresher, for example, becomes more human than the workers who feed it: "The hum of the thresher, which prevented speech, increased to a raving whenever the supply of corn fell short of the regular quantity."[39] The thresher prevents the workers from speaking, the ultimate signifier of humanity, but itself raves when it needs to be fed, its mouth agape and shrieking.

A synecdochical machine is a loudmouth that visually swallows human sacrifices in the film *Metropolis*. The Moloch machine physically destroys the selves of the laboring poor who worked and died under the dystopian city. The Moloch scene is prefaced by music that invokes the sound of tribal drums, warning audiences in 1927 that they were about to turn toward a primitive world. As viewers enter the room with dozens of giant machines, they see that the workers' movements are carefully synchronized to the soundtrack, the instruments visually imitating a new, steady, mechanical beat. Gradually, additional layers of sound are added, at first imitating the sound of the belts that drive machines, then rising and quickening to signify increasing pressure in the steam boilers and the increasing speed of modernity until finally the machine explodes in a crescendo of timpani and rising violins that mimic blasts of upward steam. The workers have failed to keep up with modern time. As they fall behind, the machine is angered, an emotion that visually turns the machine into a fuming head with a gaping mouth (see fig. 3.1). The mouth must be fed. As the machine turns into the massive face of an angry god, the sound returns, changing from a uniform, rapid mechanical beat to a slower, "primitive" rhythm. At first, the machine is fed the bodies of slaves who are dressed in ragged loin cloths, a reference to Old Testament Joshuas. The slaves do not go into the mouth willingly but instead resist the captors, who drag them toward the maw. Once the slaves have been eaten, more bodies appear, and again the sound changes, this time to a modern theme and a more rapid pace. The next bodies to be sacrificed are those of the factory workers. These men are dressed in uniforms and march in timely lockstep, without provocation, to their deaths. Satiated, the god returns to its original form, a giant machine. The machinery of modernity, personified into a towering god, has triumphed completely over time and swallowed the selves of the men who fell behind modern rhythms.

MACHINE MOUTH · 57

Figure 3.1: "Molloch" from *Metropolis*, Fritz Lang and Thea von Harbou, director and writer (1927), © Friedrich-Wilhelm-Murnau-Stiftung. Transit Film GmbH, distributor. Babelsburg: UFA.

Following the philosopher Henri Bergson, pre–World War I composers and artists embraced the shattering, transpiercing, and deterritorialization produced by modern noise.[40] The cubist painters Pablo Picasso and Georges Braque invented a multiperspectival style that fragmented individuals into hundreds of selves. Braque's *Woman with a Guitar* (1913) is in pieces. There is no single self here but rather hundreds of them projecting from as many different angles. It is not that the self is unknowable according to Braque, rather the self has been penetrated, shattered, and multiplied. If music emerges from the guitar, and given the contemplative looks on the woman's faces, one assumes that it does, then the sounds emerge in sharps and staccatos. Just as the body is fragmented by modern time and scientific discovery, the self is penetrated in *Woman with a Guitar* by sharp musical lines that don't just cut across her; these lines cut through her. The staves of the background sheet music cut the figure into pieces as if these are razor-sharp wires. Or perhaps the background lines reiterate the steel strings of the guitar. Pablo Picasso's similar *Ma Jolie* (also known as *Woman with a Guitar*), painted in 1911 and 1912, is even more fragmented. The hundreds of fragments in Picasso's image are almost unrecognizable as a human.

Likewise, the Futurists employed art and sound to fragment and penetrate.[41] Led by the poet Filippo Tommaso Marinetti, the painter Umberto Boccioni, and the composer Luigi Russolo, the Futurists envisioned a utopian world filled with the sights and sounds of machines. And they promised to challenge the borders of the senses in a smorgasbord of synesthesia. Carlo Carra, for example, challenged visual artists to render "sounds, noises and smells in painting" through the use of movement, vibration, color, "imagination without strings, words-in-freedom, the systematic use of onomatopoeia, antigraceful music without rhythmic quadrature, and the art of noises." Recognizing that words-in-freedom are sonic and dynamic rather than visual and static, Carra understood that sounds are always freed from their origins. Unlike the printed word that can never leave the page, sound enables humans to free themselves from the tyranny of the musical scale, the written edict, and the dead hand of the past. On the canvas, sounds, noises, and smells would be translated through a variety of visual devices like "the clash of acute angles, which we have already called angles of will[,] oblique lines which fall on the observer like so many bolts from the blue, along with lines of depth" and "[p]erspectives obtained not as the objectivity of distances but as a subjective interpenetration of hard and soft, sharp and dull forms."[42] Thus, Futurism is typified by force-lines, "the beginnings or prolongations of the rhythms which these very objects impress upon our sensibility."[43] For Boccioni, the rhythmic lines described by his friend Carra not only impressed, they cut through: "[A]n object moving at speed (a train, a car, a bicycle) appears in pure sensation in the form of an emotional ambience, which takes the form of horizontal penetrations at an acute angle."[44] Boccioni's *Noise of the Street Enters the House* (see fig. 3.2), for example, applies force lines and in doing so exemplifies the penetration of the self by modernity and its noisy machines. Boccioni called the painting a diagram of sound and claimed, "[T]he noises of the street rush in at the same time as the movement and reality of the objects outside."[45] In the painting, a cacophony is made by the machines, horses, and laborers that toil beneath a woman, who looks out from her balcony. But rather than being a view of the world through the eyes, things are made distant and close through aural interpretation. Thus the loudest objects are brought near and made invasive as if the auditor's physical body was forcibly opened to the street noises. Carra followed Boccioni's work with the image of another woman on a balcony, *Woman at a Window* (1912), who

MACHINE MOUTH · 59

Figure 3.2: Umberto Boccioni, *The Noise of the Street Enters the House* (1911).

is so fragmented and so penetrated by the outside world that, like Picasso's *Woman with a Guitar*, she is unrecognizable as a human.[46]

The Futurists emphasized the communicative power of onomatopoeia because they believed this to be the basis of all understanding. Indeed, the Futurists only thought to imitate a society that was already busy incorporating the sounds of modernity into its lexicons; words like *chug* (1866), *zipper* (1923), and *beeping* (1929), for example, became common in the English language.[47] Marinetti's "Zong Toomb Toomb" (1914) is a verbal manifestation of the Futurists' use of onomatopoeia. The work is based in part on poems that he produced immediately after experiencing the First Balkan War (1912–13) as a press correspondent. "Zong Toomb Toomb" remains the most famous of these poems and is a feast of noise and sound:

> . . . taratatatatata machine guns cry writhe under 1,000 bites blows traak-traak thrashes lashes pik-pok-poom-toomb juggling clowns' leaps in mid-air 200 m. high it's gunfire Down below bogs guffaws   laughs buffalo carts goads horses stamping caissons splish splash zong-zong-shaaak-shaaak rearing pirouettes pata-traak spattering manes whinnying eeeeeeeee hubbhub jingling 3 Bulgarian battalions on the march kroook-kraaak (SLOWLY DOUBLE TIME) Shumi

> Maritsa o Karvavena officers' cries clash copper plates pom here (QUICK) pok there boom-pom-pom-pom-pom here there . . .[48]

Marinetti explains his resort to neologisms that imitate the sounds of the machine age, "We Futurists initiate the constant, audacious use of onomatopoeia. . . . For instance, my 'Adrianople Siege-Orchestra' and my 'Battle Weight + Smell' required many onomatopoetic harmonies[,] always with the aim of giving the greatest number of vibrations and a deeper synthesis of life."[49] For Marinetti, to be effective, poetry could not be composed of pretty words and subtle ideas but rather the sonic equivalent of *enargia*, vivid and immediate sounds that invoke memories, emotions, frustrations, anxieties, visions, and relations. Sound was the key to understanding and controlling modernity, not to mention winning wars.

Though the original war poems were written in French, the 1926 premier and 1935 recording of Marinetti performing "La Battaglia di Adrianopoli" were spoken in Italian.[50] Because Marinetti devotes so much of the poem to onomatopoeia, the recording is at least partially legible to any listener who is familiar with the sounds of the machine age in general and of mechanized war in particular.[51] In a gravelly, masculine voice, the poet barks out a sound near the beginning of the recording that mimics a motorized vehicle. The war between Turkey and the Balkan Alliance of Greece, Montenegro, Serbia, and Bulgaria in 1912–13 was one of the first to use automobiles and airplanes. Not surprising, one can also hear in the recitation the sound of an engine starting up and, later, dying out. Soon after the beginning, Marinetti's voice imitates far-off canons and explosions. The poet never settles into any kind of rhythm, preferring to stretch out down notes (when he assumes the role of storyteller) and speed up his cadence during particularly exciting passages or when he imitates the dashes and dots of the telegraph with which he reported his stories back to France.[52] He can also be heard employing the telephone. If nothing else is understandable here, the repeated "rat-a-tat-tat" is perfectly clear as Marinetti explains machine-gun attacks against the Turkish defenses at Adrianopoli while providing the listener with an account of the sound of automated gunfire. The pounding of close-falling cannon fire is also "apparent" from the sound of Marinetti's voice as he projects in a manner that reiterates the vibrations he must have felt in his bunker. Even the passing bullets can be made out in Marinetti's voice. The battle is occasionally marked by periods of silence as the guns fall still, scouts reconnoiter, and medics gather the dead and wounded. The military marches, patriotic songs sung by the soldiers, and officers' orders are also readily apparent even as soldiers and officers are shot at by the Turks. For Marinetti, the ear's grasp of reality becomes more expansive and clear than the particularized grasp provided by the singly focused eye. The ear can hear simultaneous sounds from every direction and can perceive events that the eyes cannot. Thus, to convey a sense of total war, one must rely more on the ear than on the eye.

For the twenty-first-century listener, what is frightening about the performance is not so much the poet's enthusiasm for the violence of war but, rather, how Marinetti's hyperbolic exaggerations anticipate the sound of Hitler's orations. Unlike the flourishes of the Shakespearian stage that hark back to a supposedly more civil age, Marinetti's flourishes are violent and intentionally disturbing. He sounds like war. Reading the sound of Hitler's voice after listening to "La Battaglia di Adrianopoli," one begins to appreciate that Germany's dictator not only sounds of war to those who came after his devastation but also must have sounded like war to at least some of those who came before. The critic Kenneth Burke was suspicious of the Furher's diction, though Burke understood that Hitler's purpose was not that of the Futurists. Rather than fragmenting the self, Hitler hoped to employ the sound of his voice, his mechanized armies, and the crowd to unify a massive group into a single body politic. Understanding Hitler's revulsion with the democratic politics of fragmentation during the Hapsburg empire, Burke explains, "such vocal diaspora, with movements that would reduce one to a disintegrated mass of fragments if he attempted to encompass the totality of its discordances" produced in Hitler a desire for unification through vocalic domination.[53] In listening to Hitler, one hears the same kind of shattering that one hears in Marinetti. But one also hears a kind of bringing together—an envelopment—that makes Hitler's voice almost unique in history. Combined with the recent invention of the loudspeaker, Hitler achieved the vocalic domination that so concerned Burke. As Hitler himself wrote in 1938, "Without the loudspeaker, we would never have conquered Germany."[54]

Hitler's desire to bathe listeners in a unifying sound is evident in his deft manipulation of applause, encouraging it when he sought consensus and cutting it when it had achieved its effect, a performance magnified by Charlie Chaplin in the *Great Dictator* (1940). The sound of a crowd—cheering in particular—creates a comforting sonic envelope.[55] In *Psychology of Radio* (1935), A. H. Cantril and G. W. Allport reported that listeners were made more comfortable laughing along with a radio comedy if they heard the sound of others laughing. Phonograph companies had already discovered that device in the 1890s, as evidenced by the addition of a cheering crowd to a political cylinder in 1892 and a laughing crowd to the end of a comedy cylinder in 1895.[56] For Cantril and Allport, studio audiences provide cues to participation and inclusion thus making listening to the radio a social act that produces a "consciousness of kind."[57] Crowds create a noise that envelope listeners with sounds that are like their own when added to it, making the crowd a protector, including listeners in a "kind" and illustrating that there is strength in numbers. Applause, Hitler understood, bathes listeners in sound, enveloping them in a unified entity, a fatherland. Simultaneously, crowd noises are also aggressive. Clapping, as Stephen Connor describes, "retains its association with violence, functioning as an emblematic display on the body of the aggressor of what may be in the offing for his victim. Clapping of hands retains its association with anger, triumph and insulting contempt."[58] One voice brought many thousands together who wrapped themselves in the sound of aggression and a felt solidarity with those who shared in the vocalic domination and in the clapping. Hitler's was the voice of the father, protector, and aggressor.

Hitler also recognized the power of the radio and film to reach audiences who could not or would not attend his massive nighttime rallies. Indeed, sound was a more effective agent of control than Nazi visual icons. Hitler noted at the early stages of his efforts at propaganda, "The sound is in my opinion much more suggestive than the image. But it has yet to be learnt to take advantage of the possibilities of radio broadcasting."[59] As Hitler and his propaganda minister Joseph Goebbels learned over the 1930s, new media technologies were potent tools in drawing the German community together. Thus, Theodor W. Adorno wrote in 1940 of the authority and fear imposed upon audiences by these technologies and of the audiences' response: obedience.[60] By 1939, the Nazis had succeeded in harnessing the power of the voice, the power of applause, the power of radio, and the power of machinery. The magazine *Zeitschrift für Musik* reported, "The actual, clandestine, powerful music of our time is still alive outside of our musical art. It resounds in the marching steps of our regiments, in the rhythm of labor, in the roaring of engines and propellers . . . in the final unison of German hearts."[61] Participants in mass events, moviegoers, and radio listeners felt a feeling of belonging together.

Of all the Nazi propaganda efforts, Leni Riefenstahl's documentary film *Triumph of the Will* remains the most famous and most persuasive.[62] Produced in conjunction with the 1934 Nazi Party Congress and released in 1935, the talkie avoids individual voices until twenty-two and a half minutes into the documentary. Until then, only three sounds can be heard: music, crowd noise, and church bells. Positive music in the form of triumphant symphonies, military marches, and folk songs is most pervasive throughout this part of the film with the sounds of crowds cheering, laughing, and chanting almost as dominant. Though shown are planes, automobiles, and individuals like Hitler mouthing words, the only other sound to be heard during the opening of the film is that of church bells peeling (at 13:45). Because the crowd noises are visibly not in sync with the crowds, we know that the inclusion of specific crowd sounds was deliberate. Even in 1934, Riefenstahl understood that certain sounds wrapped viewer-listeners in a protective cocoon. When we finally hear an individual speaking, the voice is accompanied by the image of Rudolph Hess in front of a microphone. Given the camera angle, it is evident that the microphone is deliberately included in the shot. Microphones, martial music, cheers, church bells, and ultimately Hitler's cunning voice make for an incredibly persuasive aural experience, one that bathed listeners in an impermeable sonorous envelope.

The Nazis had succeeded in turning the shattering sounds of machine noise into "music" on a scale so massive that it enveloped all "German" hearts. The Futurists had reveled in the penetrating desubjectivization of sound, its cacophony repeating in the shattering of selves. Hitler believed that his subjects would best be persuaded by the sounds of a national war that enveloped them, making them one, whole Aryan people.

Sonorous Envelopes

Rather than ameliorating the noise pollution that threatened soldiers, city dwellers, and factory workers, Futurists and Hitlerites embraced the noises as symbols of modernity and progress. Not being a German, Chaplin interpreted Hitler's words for what they were—the sounds of unification and violence. In *The Great Dictator*, Chaplin plays Adenoid Hynkel, a spoof of Hitler. Chaplin is particularly effective when he satirizes Hitler's oratory in a brilliant pig-German that employs English words to make ridiculous the despot's bombast. Toward the end of Hynkel's gibberish speech, Chaplin pours water in his ear, a metaphor perhaps for Americans' unwillingness to listen closely to Hitler's dangerous rhetoric and a critique mirroring Burke's essay. As a corrective, Chaplin closes *The Great Dictator* with an oration presented in the plain style and in English. After celebrating the good earth, Chaplin laments that we have lost the way: "We have developed speed but we have shut ourselves in. Machinery that gives us abundance has left us in want. . . . More than machinery, we need humanity. . . . [Soldiers,] don't give yourselves to these unnatural men, machine-men with machine-minds and machine-hearts. You are not machines." Rather than become machines, Chaplin hoped that modern men and women would exert control over machines and celebrate the new devices' contributions to humanity.[63]

Visual artists of the interwar period led a shift in the perception of machine noise from shattering to comforting. Perhaps necessarily, the visual shift from the fragmenting lines of Futurism to the curvilinear lines of the "return to order" that art witnessed after World War I is abrupt.[64] And just as Europeans had gone to extreme lengths to celebrate the soul-shattering noises of modernity, it was also the Europeans who responded most forcefully in reaction against the celebratory sounds and images. In 1927, for example, artist George Schrimpf revisited the balcony images that Boccioni and Carra had painted in 1911 and 1912. Unlike the fragmented and penetrated women who stand on their Italian balconies, Schrimpf's women remain whole, solid, and unfragmented. Although they are not confronted with the building of a factory in their front yard, the other sounds of modernity—engines, whistles, drones—do not pierce these women. They are recognizable to us, both visually and psychologically, as entire persons and whole selves. Like them, we have grown accustomed to modernity. We have adapted to and adopted the new soundscape. Visually, Schrimpf's painting is less urban, recalling the premodernist world when eyes were not assaulted by sharp angles. That Schrimpf chose this setting at a time when he was working with Carra suggests the painting is a tribute to the older artist and illustrates that not only had their ears adjusted but so, too, had their eyes.

Modernity and its machines could have continued to be depicted as sharp edged and penetrating, but reacting against the prewar artists, Cyril E. Power and Sybil Andrews described one British Futurist's style: "To express the revolting horridness and the noisy incessant vibrating rattle, he has used ugly sharp, angular and square shapes and drabby crude colour which hurt to look at."[65] The 1920s experienced a return to order, as a result of which surrealists, dadaists, and Bauhaus artists no longer depicted humans as fragmented and penetrated by machine noise. Not only did art return to order but so, too, did the self. Where Braque and Boccioni had cut apart individuals with sharp, heavy, and angled lines, their artistic descendants did not. Remarkably, many artists during the interwar period continued to portray machinery, but rather than employing sharp lines to illustrate noise, dynamism, and penetration, these artists resorted to curvilinear lines that suggest an embrace. Machine noise under their hands no longer destroys the self, rather it envelops the self, protecting individuals against forces outside of the sound.

Thus, the noises of modernity were reshaped into comforting sounds and curvilinear lines. After World War I, composers Hanns Eisler and John Cage turned to modern noises as a complement to musical composition, not as a competitor. Perhaps alluding to Boccioni's painting *Noise of the Street Enters the House,* Eisler taught students, "When you are composing and you open the window, remember that the noise of the street is not mere noise, but is made by man."[66] In the same vein, Cage discussed a lesson for his students: "I had the lights turned out and the windows open. I advised everybody to put on their overcoats and listen for half an hour to the sounds that came in through the window, and then to add to them."[67] Rather than interpret these noises as penetrating and shattering, Eisler and Cage incorporated these sounds into a music that had the potential at least to wrap its audience in a sonorous envelope.

That return to order was reiterated in visual art. Power and Andrews, partners for much of the interwar period, exemplify the shift. These artists not only worked together on prints and paintings but they also studied music together, though not the sharp-edged pre–World War I symphonies of Stravinsky and Russolo—rather the softer sounds of baroque composers like Johann Sebastian Bach. Unlike Braque, who portrayed the guitar as a sharp instrument that cut into and through the woman who played it, Power visualized sound as a comforting envelope. His linotype *The Concerto* features curvilinear lines and soft colors. Even the sharp bows of the violins and basses are curved. Indeed, the entire stage has been curved by the performance, sonically enveloping the players. Musicians form half circles around the maestro, and their sounds are carried away from the stage in rounded waves before the notes reach the first row of the audience. The conductor visually reenacts the sound waves in his very shape, and we are comforted rather than confronted by the image.

Incredibly, his partner Andrews's *Sledgehammers* turns the sound of factory stamping into a sonic cradle that embraces a modern work gang who pound the metals that emerge from an invisible forge (see fig. 3.3). Hammers swing in rhythmic, circular patterns, the sounds emerging to wrap around the men in curvilinear lines, the greens only emphasizing the benignity of the noise, the soft reds to signal a reflection of the forge. Even the workers have been made curvy by the noise they make. Unlike the bright yellows and reds the Futurists employed, Andrews's earthly colors suggest that the scene is natural. The men in this image produce a noise like the peeling of church bells that propels the outside away, making their community whole

Figure 3.3: Sybil Andrews, *Sledgehammers* (1933). Color Linocut © 1933 Museum of Fine Arts, Boston.

and safe as the circular lines ripple outward from the center of the image. Though ostensibly about men participating in a masculine activity, Andrews's image evokes the idea of a womb that protects and keeps at bay the outside. We know that the sound of repeatedly ringing sledgehammers can be unpleasant, yet this image does not repel us away from the noise. Rather, the curvilinear and concentric lines domesticate the machine age and welcome us inside the linotype. Perhaps, it should not surprise us that the year after Andrews crafted *Sledgehammers,* Jesse Bradley and inmates from the Texas State Penitentiary at Huntsville turned the sledgehammering of a chain gang constructing a railroad into a rhythmic song that bent the sound produced by their tools into a musical phrase: "Hammer Ring."[68] That hammers ring is yet another reference to the church bells that constituted communities and comforted listeners.

Just as visual art experienced a "return to order" in the 1920s, so, too, did the sonic arts witness a "stabilization of music" led by Stravinsky, who returned to neoclassical forms of composition. This reversion brought back preromantic elements while moving away from the harsh dissonances and thought-defying rhythms of the prewar composers.[69] The extremism of prewar visual and sonic composition, the celebration of all things modern, and the most horrifying war yet experienced may have propelled the return to order in the visual and sonic arts while turning the attention of survivors to what they heard rather than what they saw. Perhaps, Marinetti had it partially correct; modernity was easier to understand through the ears than through the eyes. Or, perhaps, ears could not be shut to the machine age. The popularization of the radio in the early 1920s, for example, concentrated scholarly attention on the ear rather than the eye. Even then, the radio was thought to envelop listeners. In *The New International Encyclopedia* of 1922, sound transmitters were equated to "radio beacons, provided with a device for projecting a revolving beam of radio waves. The beams may be flashes produced in accordance with a system, just as in a lighthouse."[70] Radio transmitters are lighthouses for the ears that promised safe haven from the storms of modernity. And those transmitters broadcast radio *waves* outwards in concentric patterns, like ripples in a pond.[71] Radio bathes listeners in a protective and comforting envelope.

The idea of comforting sounds that protect the self is derived from French psychoanalysis. Didier Anzieu, a student of Jacques Lacan, recognized the phenomenon when he described the sonorous envelope.[72] Primordially, the sonorous envelope exists in the womb as the future child listens to and feels his or her mother's heartbeat and voice. The attachment to the voice that protects, however, does not end with birth as the newborn incorporates exterior sounds—the parents' voices, sounds of the home, music—into his or her sonorous envelope.[73] The ability to adapt to new noises in a manner that recomposes the sonorous envelope makes possible the adoption of machine noises into a soundscape that comforts and protects. That sonorous envelope shields against outside forces and enables the person within to constitute a unified and impenetrable self.

The sonorous envelope that surrounds each individual protects against the shattering effects of alien noises.[74] But those same industrial noises, and sometimes voices, may also be perceived as penetrating the individual and shattering the self. Anzieu writes of the monotonic, metallic, and husky voice of those mothers who have schizophrenic children: "Such a voice disturbs the constitution of the Ego: the sound bath no longer envelops, but has become unpleasant . . . ; it contains holes and causes them."[75] If the metallic and monotonic sound of a mother's voice is perceived as disturbing and unpleasant, similar sounds made by the screaming, whining, and humming of industrial machinery are that much more dangerous to the ego. Yet, one's own voice becomes a device that people use to build a sonorous envelope for protection as evidenced in the scream of the torture victim, the mumblings of the mentally ill, the yell of a suicidal military charge, the raised voices of arguers, and the chanting of protesters. Harvard polymath Elaine Scarry describes the scream of the torture victim as prenatal and even prelinguistic. It is a primitive sound designed to protect against outside forces.[76] In their book about capitalism and schizophrenia, Deleuze and Guattari similarly describe the frightened child who "comforts himself by singing under his breath."[77] When we believe our selves are threatened, we sometimes surround ourselves by projecting our voice outwards as if it were a physical force that might protect against electroshock, CIA mindreading, bullets, demands, and arrest. It is why Melville, Davis, Dickens, and Hardy worried over the drowning out of human voices. Without those voices to protect against the machine noises, the selves of workers were penetrated and shattered. In a 1937 essay titled "On the Assembly Line," a writer for *The Atlantic* describes how workers protested the assignment of overtime on the noisy factory floor. One of the workmen explained, "We take to hollering to build up a morale which will help us lick the last hour."[78] This is the sonorous envelope.

Humans are capable of adapting to noise, whether by contesting strange and loud sounds with their own voices or accommodating noise in such a way that it is internalized. Already in 1915, writer D. H. Lawrence recognized that machine noises like those made by the train began as penetrating but later become a balm: "from beyond the now familiar embankment came the rhythmic run of the winding engines, startling at first, but afterwards a narcotic to the brain."[79] Humans could grow used to the noise the more they were exposed to it, and so rather than fearing the raving voice of the machine-mouth, people would come to rely on its constant presence as a comforting element of the soundscape of their home environment.[80] With modernization, soundscapes became increasingly composed of engine hums, metal stamping, rail clicking, whistle-blowing, clock ticking, and other noises, though these took some getting used to before becoming familiar and homey. In 1918, for example, writer Thomas Mann compared a soundscape that featured the noises of a locomotive factory to ocean waves: "We are encompassed with a roaring like that of the sea. . . . [T]here is a locomotive foundry a little way downstream. Its premises have been lately enlarged to meet increased demands. . . . Beautiful glittering new engines roll to and fro on trial runs; a steam whistle emits wailing head-tones from time to time; muffled thunderings of unspecified origin shatter the air."[81] For Mann, the comforting, ocean wave–like roar of the machines might still become the penetrating noise of modernity that shattered air and self. Like Lawrence, Mann recognized that humans could become accustomed to the noises of the machine age though it might take time. Thus, for Andrews, even sledgehammers do not necessarily produce penetrating, self-destroying sounds but rather can be incorporated into a sonorous envelope. For Mann, the locomotive factory only occasionally emits noises that threaten to render the protagonist's soul. The usual sounds comfort the protagonist in a sonorous bath. That sonorous envelope protects against outside forces and enables the person at the center of the envelope to constitute a unified and impenetrable self, established in part on the sounds that protect him or her—the familiar soundscape.

Radio sounds are easier than machine-noises to incorporate into sonorous envelopes. Through controlling volume, tuning to stations that reflect one's identity, and associating the radio with comforting locations like homes and churches, the radio protects selves from outside noises.[82] And just as radio performed as a center of sound waves that wrapped listeners in a sonorous bath, so, too, did music, particularly when broadcast over the radio.[83] Classical composers beginning in the late nineteenth century had already begun to recognize the possibility of enveloping audiences in sound by producing louder and louder symphonies. In his 1941 essay "Radio Symphony," Adorno notes, "The power of a symphony to 'absorb' its parts into the organized whole depends, in part, on the sound volume. Only if the sound is 'larger,' as it were, than the individual so as to enable him to 'enter' the door of the sound as he would enter through the door of a cathedral may he really become aware of the possibility of merging with the totality. . . . This largeness of sound has nothing to do with noisiness, but simply with the necessity of enclosing the listener."[84] The absorption and enclosing of listeners in a cathedral of sound—a sonorous envelope— enables individuals to enter into the music, a place of solace and comfort simultaneous to being a place that reassures and reiterates identity and self. Notable, Adorno distinguishes the cathedral of sound from noisiness. The one enclosed, the other disturbed.

Not surprising, composers attempted to incorporate penetrating industrial noises into music in a manner that domesticated these sounds, internalized modernity, and prevented the dismemberment of the self in favor of creating sonorous envelopes. By turning penetrating noises into enveloping music, composers, performers, and radio engineers encouraged listeners to internalize modernity and find comfort in what had once been perceived as shattering sounds. Audiences responded by gravitating toward specific musical genres like jazz and blues that adopted industrial noise and reified their identities through the protective shelter of the sonorous envelope. Music has often acted as sonorous protection against the outside but never more so than when people needed to adjust to ticking, beeping, clanging, pumping, hammering, humming, and other industrial noises. One noise in particular, the locomotive noise, became the ideal vehicle for meliorating the sounds of modernity.

Though Mumford thought the clock, which had been incorporated into music through the metronome, to be the synecdoche of modernity, he recognized that others described the locomotive as the exemplar. Walt Whitman, for example, described the locomotive as the "type of the modern—emblem of motion and power—pulse of the continent."[85] Indeed, without the railroad, the need for chronological precision would have been significantly delayed. It was both the locomotive and the clock that became the focus of American composers' and listeners' efforts to domesticate modernity. The noise of the clock is disturbing in interwar musical productions like the accompanying compositions for *Metropolis* and the opening scene of Charlie Chaplin's *Modern Times* (1936), though less so in postwar efforts like Doris Day's "Tic, Tic, Tic" (1949). Locomotive sounds, on the other hand, were adapted into music much earlier and much more smoothly.

Blues musicians have for a century employed onomatopoeia to incorporate the sounds of the railroad into the American musical vocabulary and thus to domesticate these sounds in a manner that creates sonorous envelopes.[86] R. Murray Schafer argues, "Of all the sounds of the Industrial Revolution, those of trains seem across time to have taken on the most attractive sentimental associations."[87] Perhaps, this is so because one can enter into the train. It is like the mother's womb; once inside, we are protected against the outside, the noise rippling away from us, penetrating those on the outside while defending those on the inside. Or, perhaps, this is so because the noises of the locomotive have been bent into musical forms, retraining listeners' ears to adapt to sounds of the machine age. Evidence of the incorporation of train motifs in recordings can be traced as far back as the 1890s when the comedian Russell Hunting incorporated railroad sounds into his humorous cylinders. In 1898, the recording *Down on the Swanee River* employed train bells and the sound of a steam engine for effect. In 1903, the blues composer W. C. Handy overheard a railroad laborer singing and playing an early version of what was published in 1914 as *Yellow Dog Blues*, the first recorded song to turn the rhythms of the locomotive into music.[88] By the 1920s, railroad sounds had been widely incorporated into America's vernacular music.

In 1940, Mississippi sharecropper and hobo Booker "Bukka" White epitomized how blues musicians employed sounds of the trains in his recording "Special Streamline," a reference to the art deco–styled train that transported black southerners from Louisiana and Tennessee to the factories of Chicago and Detroit. The New Orleans Streamline was responsible for physically fragmenting black communities and sonically shattering selves.[89] Indeed, it was an early version of this train that took Richard Wright from Mississippi to Chicago in 1925, a migration that would leave Bigger Thomas separated from his home soundscape and in fragments of his own. Penetrating into the ears of composers like White as noise, however, the sounds reemerged as music that comforted and enveloped.[90]

Unlike many other recording artists who had set their railroad sounds to wax before 1940, White explicitly pairs locomotive noises with musical formulations throughout the story he tells in "Special Streamline." He introduces each train motif, played on his guitar, with verbal cues: "blowin' like this," "the bell begin to toll like this . . . Make a sound like a church bell toll," "she blow'd and throw'd on the airbreaks . . . Airbreaks!" "strikin' the double iron like this," and "hear her when she squalls." Thus White keeps time by percussively imitating the clickity-clack, mimics the train's bell with the guitar's harmonics, and bends the strings to imitate air brakes. After the prompt "she squalled in," White also imitates the redshift of the Doppler effect of a moving train, during which the pitch descends as the engine passes by the listener. For the remainder of this cante fable, when White's guitar is not prompted by his voice, it continues the clickity-clack rhythm of steel on steel. These techniques—percussive guitar, harmonics, bending strings—domesticate the locomotive into a musical form that is more reminiscent of the lullaby than of the transpiercing noises of modernity.

Notable, White mostly speaks this song like he was telling a story. Similar to the mother's voice that simultaneously protects and enlightens, White's guitar playing and storytelling create a sonorous envelope and teach listeners how to turn locomotive and other modern noises into comforting soundscapes. The story describes the black experience of leaving an agrarian and preindustrial south for better opportunities elsewhere. Its protagonist, a young woman who can no longer stay with her Mississippi boyfriend because he is too poor to feed her, must take the train from Memphis, sorry to leave but having no choice. Reiterating Bigger's fight against industrial time that he could never understand, the young woman's boyfriend reveals that he, too, has not been trained to internalize the sounds of time:

> She said, "Daddy, is that my train?"
> I say, "I ain't keepin' up with the train time
> I'm tryin'-a make a few dimes."

The "I" here cannot adapt to time. Like Bigger, White had murdered a man, though the composer had been released from prison. Unlike Bigger, who was dragged away to the industrial north, White claims he will continue to scrape by on the few dimes that he earns in Mississippi, even as he fails to keep up with the train time.

Illustrating how the sound of the train has shattered the narrator, the bluesman employs many different voices in the song. The shift in perspective from narrator to young woman to boyfriend to passerby (not to mention from storyteller to singer) creates a number of "I"s, all of which are White. The sound of the train has penetrated and shattered the author into many selves. Ancient Greeks had a term for this effect, *dialogismus*, which means to make one's own voice that of another; a sort of reverse ventriloquism by which the self is spoken by a second (and third and fourth) self. Thus, White's story rhetorically performs his sonic fears about the locomotive. His body speaks with the voice of many selves, all of them him. Like Boccioni's woman on a balcony, White has been shattered by the noises of modernity.

The young woman is also at risk of being shattered. Though she makes the leap to modernity by boarding the train, she is understandably worried that the machine will swallow her like the machines described by Melville, Davis, Dickens, and Hardy:

> Hey, dad I don' wanna leave
> I believe I'll lose my mind.

Like Whitman, who described trains as "fierce-throated," and the workers in *Metropolis* who were eaten by Moloch, the young woman is threatened by the machine.[91] Will she be shattered by the penetrating noise of the locomotive and become a murderer? In sympathy with the young woman, the listener takes the same leap by listening to White's storytelling and to the bending of machine sounds into a melody. Because the noises have been transformed into music, the song protects rather than shatters.

Blues artists like White bent the sound of the train into music. So, too, did many jazz composers. In a study of the 1930s, Dinerstein recognized that big-band swing artists made sense of factory noise by turning machinic rhythms into propulsive beats.[92] Like the jazz composers, White recognized a need, whether his own or that of Mississippians, to make sense of the sound of the train in 1940. Twenty-three years later, he rerecorded "Special Streamline" as "New Orleans Streamline." By the second recording, however, White had internalized industrial time; he had been drafted into the navy during World War II and later worked at a tank factory in West Memphis, Arkansas. Thus, the story in the second version is different from the first. Now we hear about a girl who left Memphis and has arrived in Detroit, home to hundreds of thousands of black migrants, many of whom worked on Ford's assembly lines. And though the guitar breaks are similar to those of the 1940 recording, White no longer feels compelled to explain them as imitations of locomotive sounds intended to domesticate external and penetrating noises. He only introduces two of the guitar riffs with: "He heard that streamline tippin' in there that morning," and "You could hear it chunkin,' rail hittin.'" Instead of prefacing each guitar break with an explanation that these were train imitations, White lets his new listeners make the connection themselves. Indeed, the listeners needed no explanation for train sounds bent to form music because they already knew.

Radio Waves

Where Bigger Thomas fought against modern, industrial noise and ultimately lost, Richard Wright and Bukka White adapted the sounds into the blues. Of course, Bigger was not the only character to fight, nor were Wright and White the only people to adapt. Indeed, the historical progression of adapting to modern noise has been universal and unstoppable. At the fringes, cities have enacted noise-pollution ordinances to limit industrial disturbances (like jackhammers) to daytime hours or that reduce the loudness of machines (like jet planes) in certain neighborhoods. However, the trajectory of the past two centuries has been one of ever-increasing mechanical noise. Our offices buzz with the sounds of fluorescent lighting, our homes whoosh with the noises of air conditioners and central heating, our yards trill with the rumbling of passing automobiles, the whirring of lawn mowers, hedge trimmers, leaf blowers, and on and on and on. Incredibly, most of us adapt to the noises and, indeed, come to expect these sounds. Unlike Mississippians in 1940, we do not need help from artists to make sense of the sounds of modernity. We largely know how to adjust.

Novelists, artists, and musicians domesticated the noises of modernity from clocks to sledgehammers to locomotives by turning those penetrating sounds into comforting and familiar music, thus training the ears of listeners to adapt to the machine age. Radio, in particular, provided a metaphor for modernism that protected people during the interwar period against the roars, groans, and clanks of the machine age. Though the concept of the sound wave had been around since the eighteenth century, those waves were not widely observable until radio enthusiasts made sense of sound for consumers in the 1920s. Those who sought to communicate radio's powers to a wide audience were fond of explaining the science of sound transmission through the use of comforting analogies, in particular, the wave. In *Radio-Telephony for Everyone* (1922), author Laurence Marsham Cockaday explains, "We now know what a sound wave is. Have you ever noticed that the seats in a theatre are invariably arranged in a semi-circle? This is done to take advantage of the fact that all waves—such as sound waves, light, heat, and also radio-waves—travel in ever-increasing circles." Anticipating British modernist Cyril Power's *Concerto*, Cockaday employs the rounded motif of the musical stage, a place from which comforting sounds ripple outward, to describe how a radio transmitter projected electrical noise. But Cockaday felt the need for another, more soothing analogy: "The transmitting station may be likened to the stone thrown into the water, and the receiving station to the corks on the surface of the water."[93] Like Mann, who had converted the machine-age noises of a locomotive factory into waves by comparing the sounds to the ocean, Cockaday writes as if we are bathed in sound waves projected by the most modern of machines just as we would be if we were to swim in the ocean, or a pond, or a womb.

The analogy of sound waves made sense to radio audiences. In 1928, a German listener wrote to his favorite radio station,

> Wave, be aware of your many shapes
> And, all embrace, weave
> At the world's wheel, entrusted from above
> The new and wider spirit of the human race.[94]

For this listener, the radio waves that surrounded him embraced and weaved at the world's circular wheel. And for him, God had entrusted this device to make peace in the world. Likewise, Adorno, in "Radio Symphony," writes that the radio had a narcotic effect that he termed its "surrounding" quality: "The drug tendency is very clear in Wagner where the mere magnitude of the sound, into whose waves the listener can dive, is one of the means of catching the listeners, quite apart from any specific musical content."[95] The music did not create the waves into which listeners dove, but instead the radio-machine and its amplitude immersed audiences in a druglike state of happiness and pleasure. Lawrence had similarly written that machine sounds were a "narcotic." Not surprising, many believed in the healing powers of the radio waves that curved around people. Doctors began to employ radios to treat patients, and patients self-healed through the invisible powers that emanated from their speakers. Dr. W. F. Jacobs extolled the healing powers of radio waves in 1922: "Radio deserves to be ranked with the best mental therapeutic agencies. In fact, for hundreds of cases the radio telephone can be prescribed as the one best treatment."[96] The sound waves that the radio emitted would, ostensibly, envelop patients, providing comfort, serenity, and other healing powers.

The adaption of machine noises to music, radio, and the metaphor of sound waves helped individuals to internalize the noises of modernity, thus creating sonorous envelopes and domestic soundscapes. Aided by musicians whose songs were broadcast by radio stations and by visual artists who made the self whole even in the face of industrial sounds, listeners during the interwar period normalized the most disturbing part of modernity: its noise. Because scholars have focused on words and images, the threat posed by those noises to the self has been largely forgotten. The study of sound helps us to recover that insecurity and to remind ourselves that the self has not always been impenetrable during the modern age.

4

The Race of Sound

Jimmie Rodgers, the father of country music, recorded a few blues tunes at the Victor Corporation's Hollywood studio during a visit to California in July 1930. One of the songs he recorded required more than he could provide through his weak voice and simple guitar picking. So a pianist and a trumpet player were drafted to accompany the country crooner in the studio. Neither of these artists' names were included on the records produced by Victor. When the single "Blue Yodel #9" was marketed, it went out under Rodgers's name alone. Those who listened to the recording from its release in 1930 until it was performed on the *Johnny Cash Show* on October 28, 1970, probably wondered who the trumpet player was because the performance was extraordinary. The question was answered in 1970 when Cash was joined by Louis Armstrong, who played the song for the first time since he had recorded it with Rodgers. Armstrong announced to the audience that he and Cash were going "to give it to 'em in black and white."[1]

That the father of (white) country music and the great (black) jazz trumpeter played together in 1930 complicates assumptions about Jim Crow laws and racial segregation during the interwar period. We believe that Jackie Robinson broke the color barrier when he played for the Brooklyn Dodgers in 1947 and that country music has always been a white musical form even as blues and hot jazz (in contrast to cool, dance, or swing jazz) have always been black. But why? In the wake of the civil rights movement of the 1960s, scholars found an origin story about racial integration in America. Searching for the answer in a period dominated by television and rock 'n' roll, those scholars overlooked radio, blues, and B movies, focusing instead on televised baseball and variety programming like *The Ed Sullivan Show*. In a culture in which sports dominated the visual airwaves and rock 'n' roll dominated the sound waves, it seems almost inevitable that scholars began their origin stories with the integration of baseball by Robinson or the appropriation of black music by Elvis Presley.[2] The scholars would have had difficulty recalling the impact of jazz personalities, radio, and the blues. But the photos of Robinson in his Brooklyn Dodgers uniform in 1947 are like the silent films of the 1920s—they tell only half of a story.

I do not wish to detract from the courage of Robinson, who withstood the racial insults of fans and players during his heroic professional baseball career. But Robinson played most of his games in urban settings. Some of the fans who saw Robinson were not watching a black man play with whites in public for the first time. Some would have already seen blacks and whites playing together in the dance halls, jazz clubs, and speakeasies of cosmopolitan and biracial cities. Indeed, black and white musicians had occasionally played together from the beginnings of jazz, a pattern repeated in other musical genres, particularly the blues. Listening to jazz and blues between 1919 and 1941 presents a different picture of integration. By studying the interwar period, a period when sound was as significant for knowing the world as sight because of the ubiquity of phonographs, telephones, and radios, scholars may be able to gain a fuller appreciation of the construction of race and racial integration in the twenty-first century.

The very terms applied to race—*black* and *white*—remind us of that construction's visual genesis. Those terms have resulted in the production of scholarship that focuses on visual aspects of race and ignores other sensory constructions. Thus, important studies on blackness and whiteness describe the sight of whites in blackface, visual passing, and black fashion.[3] Winthrop D. Jordan's seminal study, in particular, emphasizes the importance of appearance in the creation of distinct racial identities, a result of his argument that race's origins are rooted in the visual observation–obsessed age of reason.[4] Studies done on the sound of race largely emphasize dialects and music as adjuncts to black and white. However, Lawrence W. Levine, in his still-seminal *Black Culture and Black Consciousness* (1977), employs sound in a nuanced manner to make sense of black identity. Yet, even Levine, who taught students to listen, does not read sound as constructing identity; he reads identity as already constructed. For Levine, sound reiterates identity.[5] Yet, scholars can discover that sound, when read deeply, also constructs identity.[6] Where Levine argues about the sound of race, I argue about the race of sound. Levine's argument is based on the idea that we can hear the race of individuals in their sounds, thus, the sound of race. But race has also been sonically constructed—never more so than during the interwar period—thus, the race of sound.

Musicologists and historians have studied the intersection of sound and race in some detail, but besides the problem that popular music has only recently become an acceptable subject of study, scholarship has begun with the assumption that music reiterates race rather than constructs race. Burton W. Peretti, a student of Levine, has written a scholarly account of the early years of jazz and its racial implications. His study as well as Kathy A. Ogren's *Jazz Revolution,* also about jazz and race during interwar America, accepts race as a given. According to these authors, whites could play "black" music, but they could not become black, and blacks could play "white" music without ever becoming white.[7] But the evidence suggests that in a kind of sonic passing, whites were perceived as black, and blacks were perceived as white when their music violated the period's expectations. Because race was left in a space of invisibility, a space where distinctions based on color were sonically constructed, when listeners heard Blind Willie Dunn and His Gin Bottle Four playing hot jazz, they knew they were enjoying black musicians. They were wrong, of course, but their false knowledge is telling about the constructed nature of race during the interwar period.

Making poor arguments about the race of sound is difficult to avoid even in the twenty-first century. Just as interwar listeners held assumptions about sound that could be manipulated and reconstructed, listeners in the twenty-first century have their own assumptions about the race of sound. In analyzing the sounds produced during a 1923 recording session that brought together the black pianist Ferdinand "Jelly Roll" Morton with a white jazz band called the New Orleans Rhythm Kings, for example, it is easy to assume that Morton was discriminated against because he was black. After all, it is difficult to hear Morton on the track "Mr. Jelly Lord." Was he intentionally snubbed by the recording engineer? Indiana was then home to about 250,000 Ku Klux Klansmen, who recorded at the same studio, so this would make sense. But, in fact, the primitive technology used at Gennett Studios was much better at picking up the sounds of brass and reed instruments than it was at picking up the piano.[8] Thus, the cornet, trombone, clarinet, tuba, and sax can be heard distinctly. The piano (not to mention the banjo) is difficult to hear on this and most jazz-band recordings cut before 1925, when microphones were introduced to recording studios. The original assumption, then, was incorrect. Morton was not discriminated against; rather, his piano is difficult to hear because of the era's recording technology.

Recording engineers tolerated if not encouraged the racial integration of jazz and blues musicians because radios and phonograph players detached sound from the visible world. In the history of race relations, performers of instruments were integrated first because listeners could not see the race of the musicians and because instruments were difficult to identify according to race. It was not uncommon for musicians to perform across racial boundaries during the Jim Crow period, though at first only in recording studios and closed-door clubs or radio stations. These sonic technologies enabled further integration because the invisibility they produced disturbed racial categories, a disturbance that contemporary sound theorists attempted to resolve.

One solution to the problem of invisibility was to make sounds visible, and photographs begin to show the racial integration of performers themselves during the 1930s. Even as jazz fans became increasingly aware of the integration of popular music and though many had witnessed this integration during live concerts, until 1937 there were no permanent markers of this seismic shift in social norms about segregation: the norm remained that white and black Americans were not supposed to be playing music with each other. Where black and white musicians had already been playing together in closed recording studios and after-hours nightclubs, they began to appear together publicly starting in 1930. During the early 1930s, these performances were instrumental only but remained partly invisible because they were not visually recorded. Memory is a funny thing. We think we see something but without a permanent record of that memory can never be entirely sure of what we saw. Though private photographs of integrated performances date to 1934, an integrated photo did not appear in print (in a publicity photograph for Brunswick Records) until 1937, leaving a gap between memory and knowledge from 1930 when audiences began to see live integrated performances to 1937 when audiences began to see photographic evidence of that integration. This created cognitive dissonance. Photographs and motion pictures began to close this gap in 1937.

The closing of this cognitive gap provided the impetus for the closing of another gap that created cognitive dissonance: the gap between the body and the voice, the latter of which was the last element of music to be integrated. Though it may seem counterintuitive, the voice more closely marks the true self than the body. Given that we believe in racial categories, we assume that the voice, not the body, is the most accurate marker of race. The invisible media of radio and the phonograph player required Americans to confront this presence of the voice and then to assign race. An analysis of presence and race leads to a discussion of the ancient rhetorical concept of *colores*, a term reenergized by speech scholars during the first half of the twentieth century. Billie Holiday sonically exemplifies the figures that rhetoricians and musicians had been practicing for centuries, particularly in her protest song "Strange Fruit." Holiday colored the beliefs of her audience through her clever use of vocal *enargia* and several sonic figures. She doubled and tripled the meanings of words through the use of her voice, making the poem she sang that much more present. The presence of the voice and vocal coloring left listeners to confront uncanny feelings about the individuals who emanated from their radios and phonograph players. *Uncanny* has a double meaning, first, something simultaneously familiar and strange (from Freudian psychoanalysis) and, second, supernatural. The two fit seamlessly as seen in *The Shadow*, discussed later.

The problem of the uncanny—that which is simultaneously familiar and strange—was resolved not by making visible, as photographs had done for black and white bodies, but by drawing attention to invisibility, particularly in serials like *The Shadow, Amos 'n' Andy*, and the *Invisible Scarlet O'Neil*. These serials drew the attention of listeners to the problem of the uncanny and in doing so helped to naturalize the invisible voices that emanated from machines. Interwar concerns about invisibility and blindness, the inability to see, exposed that part of the normally hidden episteme concerned with racial distinctions. Because we are able to read sound, we are able to understand how sound constructed race.

Instruments of Integration

Until 1919, black Americans were almost never heard absent their bodies. But thanks to a series of U.S. district and appellate court rulings, culminating in *Victor Talking Machine Co. v. Starr Piano Co.*, the patents that had permitted three companies (Victor, Columbia Phonograph Company, and Thomas Edison's National Phonograph Company) to monopolize the recording industry were found abandoned, thus enabling small recording companies like Starr Piano's Gennett label to enter the market. This resulted in startup companies filling the demands of niche markets for musical acts of all kinds, including jazz, blues, zydeco, and "old-timey" performers.[9] The first black musical artist to be recorded was Mamie Smith, who cut two songs for Okeh Records in 1920. Smith's blues recordings sold so well to black consumers that most of the recording companies quickly developed catalogues of "race records," a term coined in 1922. Because of the competition engendered by this very free market after the *Starr Piano* decision, labels like Okeh and Gennett situated themselves at the cutting edge of music, recording hundreds of acts that otherwise would not have garnered attention. Where Okeh recorded popular black blues beginning in 1920, Gennett was one of the first to record jazz performed by black musicians, most notably that of King Oliver's Creole Jazz Band, a band that featured the young trumpeter from New Orleans who would soon record with Rodgers.

Armstrong had gotten his professional start with Fate Marable's bands on riverboats that plied the Mississippi River. Marable, a black pianist, led integrated bands in the liminal and quasi-lawless space between land and water between 1908 and 1917, hearkening back to mixed bands in nineteenth-century New Orleans.[10] The Jim Crow–like segregation of musicians became the norm in that city only at the end of the nineteenth century.[11] Though integrated, live performances sometimes occurred on the waters of the Mississippi, black and white musicians did not record together until 1923 when Jelly Roll Morton accompanied the seven-piece New Orleans Rhythm Kings.[12] Even after the lawsuits were settled and the technology improved (most significantly by the introduction of the microphone, a device borrowed from radio in 1925), recording sessions were not the modern, complicated multiweek affairs with multiple takes, dozens of tracks, and overdubbing that they have become today. In the 1920s and 1930s, musicians would sometimes cut as many as a dozen songs in a single session. And every musician who performed on a song played live with every other musician as the recording devices only created one track (a literal word referring to the groove etched into wax cylinders). Songs were often performed and recorded more than once, presenting opportunities to change band lineups for various reasons including sound quality and concerns about racial integration.

After 1927, recordings by black and white musicians in almost every genre became common. Teddy Wilson, a black jazz pianist, notes, "In those days there was no racial integration in jazz orchestras in public, although there was on jazz recordings, where it didn't matter."[13] It didn't matter at these recording sessions because the musicians could not be seen and because audiences would have assumed that the bands were segregated. Rather than rely on instruments to determine the race of sound, listeners fell back on assumptions about musical genres and racial segregation. From the "old timey" recordings by Gennett and Victor in 1927 to blues and zydeco recordings by Columbia and Okeh after 1928, the history of recorded music during the interwar period illustrates many instances of racial integration.

Though instances of behind-closed-doors integration were increasingly common, music and broadcasting corporations in the 1920s attempted to impose and affirm this period's expectations about the sound of race and the race of sound. Most American musical forms already had an aura of race, especially jazz and blues. Recording companies attempted to solidify this racialized sound through a variety of methods. For example, to preserve the "purity" of the artists' races, music corporations sometimes did not release recordings cut by white and black musicians performing together. When Jimmie Rodgers cut "Let Me Be Your Sidetrack" in 1931, for example, he performed the song at least twice. The version that Victor released was a recording of Rodgers accompanied by his own guitar. But Rodgers had also performed that song with Clifford Gibson, a black guitar player. That track was archived by Victor and not released until 1992. The second, long-unreleased recording is musically superior to the released version, suggesting that Victor may have wished to maintain Rodgers's whiteness by preventing consumers from discovering that he played with black musicians.[14] The music corporations also erased the names of white musicians from recordings with black musicians as in the country-blues recording by Rodgers and Armstrong, the jazz records of Charles Creath and Sonny Lee, and the zydeco records of Amédé Ardoin and Dennis McGee.[15]

Assumptions about jazz led to many kinds of racial manipulations. Writing in 1922 for *The Atlantic,* Carl Engel illustrates a few of these assumptions: "The term jazz, as applied to music, is rather elastic. It embraces not only the noisy-noisome sort, the jumble-jungle kind, but a type that refines upon and meliorates the racy stuff of wilder species with matter of a distinctly and engagingly musical nature."[16] Engel's ideas were common suppositions by whites about the sound of blackness, an assumption that recording executives enforced. In the 1920s, Fletcher Henderson's Orchestra (a black band that played soft dance music) was usually forced by studio executives to record hot jazz, a noisy-noisome form of music that the orchestra was not comfortable playing. Conversely, Jean Goldkette's Orchestra, featuring some of the most adventurous white jazzmen of the interwar period, was almost always forced to play slow dance music.[17] Similarly, Benny Goodman had his bands play in a light, smooth style when he cut his earliest records.[18] Goodman worried that decisions to record and perform "black" music might damage his career. Perhaps he had heard that a jazz band named the Wolverines were called "white niggers" because they played hot jazz.[19]

Another method employed by the recording industry to ossify racial categories was still more subversive. Labels set expectations—black bands played hot jazz, and white bands played slower numbers—and then fulfilled them. Because these companies controlled all of the information about the recordings—the information provided was usually limited to the name(s) of the performer(s) and the composer(s) of the tune—the market could be manipulated through false advertising and aliases. Gennett, for example, recorded the white group Memphis Five, under contract to another label, as Ladd's Black Aces both to circumvent the exclusive recording contract and to build its race-records catalogue. Billed as Ladd's Black Aces, the Memphis Five were sonically transformed into a black band because Americans assumed that no members of a white band would call themselves "Black Aces" and because they played hot jazz.[20] Wilson recalled recording with an integrated band named the Chocolate Dandies, another phrase that invoked race. By taking advantage of assumptions about words like *black* and *chocolate* and by enforcing assumptions about hot and dance jazz, record companies manipulated Americans into categorizing music according to race and in the doing solidified a sound vernacular that *colored* the race of musical artists.[21]

The record companies had tremendous power to construct racial territories through the phonograph and radio. Yet, a lurking danger rested in the invisibility of players and audiences. Invisibility was widely perceived as sonic technology's haunting side effect that had the potential to disturb all kinds of identity constructions, even between the living and the dead.[22] This disturbance remains visible in archives, in novels about sincerity, and in arguments waged by theorists over the invisibility of sound. From the beginnings of Edison's phonograph, journalists responded to press releases about that invention by emphasizing the last possibility described: that of keeping the voices of the dead.[23] Similarly, in the 1890s, listeners compared a recording of recently deceased Robert Browning's voice to a "séance."[24] Likewise when the radio became practicable, writers William James and Rudyard Kipling compared the device to the séance.[25] Psychic metaphors like "telepathy," "magic," "dream," "ghost," "mediums," and "communicating with the dead" became common in discussions about the new sound technologies.[26] The disembodied sounds produced by radios, phonographs, and telephones disturbed categories like "living" and "dead," not to mention "race," that we normally take for granted.

Invisibility disturbed because it had the power to reveal that which no one wanted revealed, secrets like those of selfhood, sincerity, and social constructions. Literary accounts of the phonograph from Bram Stoker's *Dracula* (1897) to Frank L. Baum's *Patchwork Girl of Oz* (1913) to Arthur Conan Doyle's "The Adventure of the Mazarin Stone" (1921) trade on the idea that the machine would reveal painful truths. In *Dracula,* Dr. Seward refuses to allow Mina Harker to listen to phonograph recordings that he has made because the recorded voice is truer than the written word. After Seward's phonograph player is destroyed, he feels as if a part of him were missing.[27] The talking phonograph in Baum's *Patchwork Girl of Oz* is also revealing. The phonograph exposes the imposition of elite culture on the masses and the insincerity of listening to symphonic recordings only to show adherence to expectations.[28] Sherlock Holmes in "The Mazarin Stone" also employs the phonograph to reveal truth. By separating his sound from his body, Holmes tricks criminals into revealing their plans and the whereabouts of a key piece of evidence.[29]

The disturbing nature of invisibility produced a debate among theorists and composers. Some, like Theodor W. Adorno, believed that music should be heard and not seen and that listeners should be trained to shut out everything but the sound.[30] His contemporary Rudolph Arnheim explained why in 1936:

> This new experience [of blindness] starts with music emerging out of the empty void. There is no one sitting in front with his instrument ready. No disparity between fifty waiting men, from the violins in the front to the kettle-drums at the back, and the one modest flute which perhaps has to start the piece all alone. The flute now sounds really as tremulously little and lost in nothingness, as was the composer's intention when he wrote the beginning as a solo. The flute plays, and no longer sounds like the isolated part of some nice man "in the act of playing" whose appearance never changes.... [O]nly the course of the single line of melody exists; all the action is pure movement. The flute is quite alone and suddenly the oboe joins in, likewise emerging from nothingness, unexpected and coming to life only at the moment the composer brings it in, not previously present in "counting bars." And so the work is gradually built up. Whoever has nothing to play vanishes completely out of the picture, simply does not exist.[31]

As if in response, Igor Stravinsky argued that same year, "I have always had a horror of listening to music with my eyes shut, with nothing for them to do. The sight of the gestures and movements of the various parts of the body producing the music is fundamentally necessary to be grasped in all its fullness. All music created or composed demands some exteriorization for the perception of the listener."[32] The year before, John Erskine had anticipated both Arnheim and Stravinsky: "When a person sees a symphony as well as hears it, some of his pleasure comes from the behavior of the performers or the graceful antics of the conductor.... But when he listens over the radio he hears every note, and nothing else."[33] Ironically, they all argued in favor of curing the problem of sonorous disembodiment: Arnheim and Adorno by making invisibility familiar, Stravinsky by making the sonic visible, and Erskine by arguing for both.

Not surprising, invisibility was a common sonic trope of the interwar period as evidenced by many blind blues players; they could not see the races of the musicians they learned from or the audiences they played to. Sonically, invisibility and blindness are two ways of trying to think about and grapple with the same effect—an inability to see what is heard. Ralph Ellison's novel *Invisible Man* (partly composed in the late 1930s), a book that is deeply sonic because of Ellison's musical training and his familiarity with New York's jazz scene, exploits the trope; white society is blind to the musical protagonist. Ellison's prefatory passage about a protest song performed by Armstrong in 1929 is striking:

> I'd like to hear five recordings of Louis Armstrong playing and singing "What Did I Do to Be So Black and Blue"—all at the same time. Sometimes now I listen to Louis while I have my favorite dessert of vanilla ice cream and sloe gin. I pour the red liquid over the white mound, watching it glisten and the vapor rising as Louis bends that military instrument into a beam of lyrical sound. Perhaps I like Louis Armstrong because he's made poetry out of being invisible. I think it must be because he's unaware that he is invisible. And my own grasp of invisibility aids me to understand his music. Once when I asked for a cigarette, some jokers gave me a reefer, which I lighted when I got home and sat listening to my phonograph. It was a strange evening. Invisibility, let me explain, gives one a slightly different sense of time, you're never quite on the beat. Sometimes you're ahead and sometimes behind. Instead of the swift and imperceptible flowing of time, you are aware of its nodes, those points where time stands still or from which it leaps ahead. And you slip into the breaks and look around. That's what you hear vaguely in Louis' music.[34]

In an essay titled "Production-Reproduction" (1922), the Bauhaus theorist Laszlo Moholy-Nagy describes a "language of the groove," which applies particularly well to Ellison's passage.[35] According to Ellison, Armstrong slips into the groove and disappears, all the while exercising the power to potentially disturb visual assumptions and in particular race.

Many interwar recording artists took advantage of the power of phonographs and radio to disturb racial categories. Two popular black artists during this period were Blind Willie McTell and Blind Willie Johnson, gospel-influenced, black bluesmen.[36] A third, "Blind Willie Dunn," was fictional, though "he" recorded five songs in April and May of 1929. This last blind

Willie illuminates the power of invisibility. Blind Willie Dunn and His Gin Bottle Four were a supergroup. Four of its members were acclaimed musicians in jazz and blues; they were two black artists, King Oliver and Lonnie Johnson, and two white musicians, Eddie Lang and Hoagy Carmichael. The band presented themselves, in their very name, as if they were black bluesmen. Blind Willie Dunn and His Gin Bottle Four did not chose that name to create record sales, as their own names were better marketing devices. Rather the band had to be blind, otherwise, society would not permit them to play together. And if their songs were to be heard on records and the radio, radio-station personnel and audiences would also have to be blind to the interracial composition of the band. Blindness as the B-side of invisibility was a trope that worked in the musicians' favor because it permitted them to play together.

Blind Willie Dunn also served as parodic social commentary. This is evidenced by the synesthetic song titles "Jet Black Blues" and "Blue Blood Blues." "Jet Black Blues" alludes to the corporate construction of blues as a pure black art form, when in reality there were plenty of white bluesmen, none more famous than Jimmie Rodgers, who was accused by one music critic in 1928 of being a "white man gone black."[37] Blind Willie Dunn and His Gin Bottle Four also performed "Blue Blood Blues," a play on the blue blood of white listeners like the subjects of F. Scott Fitzgerald's short story "Dice, Brass-knuckles, & Guitar" who pay to learn black dialect and dance in black jazz clubs. In both songs, the musicians emphasize the parody by giving the scat vocal not to Johnson, an accomplished black singer, or to Oliver, who trained Armstrong (the most famous of all scatters), but instead to Carmichael, a fine composer and tolerable pianist but a lousy singer. The simple act of playing together informs us that Oliver, Johnson, Lang, and Carmichael were taking on racial prohibitions. The musicians in this band did not just take on visible race codes; they also sonically challenged those constructions. This was possible because audiences were blind to the musicians and their instruments. Because of invisibility, instrumental performance became racially integrated in the 1920s.

Visible Bodies

Blindness as used by Oliver, Johnson, Lang, and Carmichael echoed invisibility as it prevented the public from seeing racial integration. Armstrong's performances illustrate another way of pairing blindness with invisibility. A white college student named Charles L. Black Jr. describes an October 12, 1931, performance in Austin, Texas: "Louis played mostly with his eyes closed; just before he closed them they seemed to have ceased to look outward, to have turned inward, to the world out of which the music was to flow."[38] Like the integrated musicians who played behind closed doors, the great jazz trumpeter closed his eyes to the public, becoming blind to the segregated music hall from which white and black youths watched from different sections. Armstrong made the audience invisible. Even so, white consumers were no longer blind to Armstrong.

Before 1930, instances of blacks and whites playing music together in closed recording studios and after hours in closed clubs were not uncommon. Most famous, Armstrong and Oliver jammed with "the Austin Hill Gang," a group of teens including Eddie Condon, Bud Freeman, and Mezz Mezzrow, three musicians who became important jazz integrationists in the 1930s. Similarly, Artie Shaw played with black artists in New York City before the integration of public performances.[39] In 1931, black and white jazz musicians began to publicly transgress Jim Crow rules. From a distance of eight decades, it is impossible to know how often whites and blacks played together, but in that year, Armstrong was arrested outside a Los Angeles jazz club for smoking marijuana with the white drummer Vic Berton after a concert.[40]

Because black and white musicians gradually became accustomed to playing together and because white audiences in cultural centers like New York, Chicago, and Los Angeles grew accustomed to the sight of racial integration, integrated bodies became increasingly visible. Until the 1930s, the absence of interracial images of recording artists supported assumptions that blacks and whites did not play together. But during the 1930s, this assumption was undermined by the inability of music and radio corporations to control how audiences completed enthymemetic arguments about the race of sound. The interwar sound vernacular, drawing upon the social knowledge of race, segregation, and the old dominance of vision, enabled Americans to perceive race in music and to make arguments about the racial composition of the sounds they heard from their phonographs and radios. But the absence of the visual also made it possible to undermine the premises upon which the arguments were founded. Young people, it appears, were particularly impervious to the racial constructions that music corporations attempted to enforce.

As the 1920s progressed, increasing numbers of young whites attended performances by black jazz musicians, a fact attested to in literature and personal accounts. In Carl Van Vechten's 1926 novel, *Nigger Heaven*, a black character rejects the idea of going to a nightclub because its clientele has too many "ofays."[41] Similarly, Van Vechten's friend Langston Hughes writes in his

1926 poem "Harlem Night Club" about whites and blacks dancing together at an integrated establishment.[42] Van Vechten and Hughes describe racially tolerant practices in New York City. The trend appeared in other urban centers. Whites attended jazz performances at Chicago's black nightclubs beginning in the early 1920s. Ellison remembers white women breaking race laws to attend an Armstrong concert in Oklahoma City, Oklahoma, in 1929.[43] But the lack of a permanent record of interracial transgression caused cognitive dissonance. This gap between the knowledge of racial integration and the belief that America was a segregated nation must have left jazz fans to wonder about cultural norms.[44] During the interwar period, a period of intense segregation when the Ku Klux Klan claimed millions of members, instances of integration were memorable. The absence of images of these events, and thus a permanent record that these had occurred, may have contributed to cognitive dissonance. Those who had heard or seen integration in a time and place where it was not supposed to happen may have doubted their own eyes and ears when more permanent evidence than memory failed to appear in print or on film.

Though a growing number of people knew that music was integrated—black and white concertgoers who were not segregated at the venue, friends who heard about such events, patrons of chic city clubs, radio- and record-company personnel, readers of Van Vechten and Hughes, the musicians themselves—no visual and public record of this integration existed until 1937, as if memories were incorrect and audiences had not witnessed the integration they thought they had experienced. This absence of the experienced is a psychologically disturbing phenomenon. Cognitive dissonance often forces the auditor to resolve the dissonance, usually through rejection of one of the factors causing the dissonance, in this case, the witnessing of integration or social expectations about segregation. Beginning in 1937, the production and distribution of images reconnected experience with memory while undermining racial constructions.

Perhaps the first image of black and white musicians playing together was taken in a recording studio. Though a testament to the courage of the musicians who participated in this interracial session and their willingness to be photographed, if only for history's sake, this 1934 image appears to illustrate the dangers of combining black and white players on film. In his professional career, Goodman cared about two things more than all else: the quality of the musicians he played with and his own success. He hesitated, for example, to play with the many black musicians he respected because he was concerned that such collaborations would hurt his record sales, particularly in the South and Midwest. Yet he had this photograph taken with session musicians, including Coleman Hawkins, a black saxophonist considered among the best at his instrument (see fig. 4.1). Where most of the musicians look at the camera, Goodman (not in the detail of the photograph) looks away. Hawkins also looks away, and his forced smile makes him appear uncomfortable. The white trumpeter, Charlie Margulis, nervously watches Hawkins as if being photographed with a black musician might also endanger Margulis's career. That was still a real worry for white musicians in 1934. Even so, Sonny Lee, the trombonist, looks confidently and comfortably at the camera, perhaps because he had been playing with black jazzmen since recording with Charles Creath in 1927. These professional jazz artists wanted to play with the best musicians, regardless of race, and found the recording studio a safe space because the product of these sessions would remain invisible to the public. But the sonic product of that 1934 gathering may reflect the presence of the photographer. As one jazz scholar notes, Hawkins "was not at his best at this session."[45]

The first photograph of a biracial group made public appears to be a 1937 image arranged by Helen Oakley, who stands to clarinetist Artie Shaw's left (see fig. 4.2). Oakley, who played a key role in the integration of jazz in Chicago in 1936 when she convinced Goodman to play with Fletcher Hen-

Figure 4.1: *Left to right:* Musicians Dick McDonough, Sonny Lee, Charlie Margulis, Mannie Klein, and Coleman Hawkins in Benny Goodman's studio (1934). (Detail of photograph 1510.1, folder 69, box 28, MSS 53, Benny Goodman Papers, Gilmore Music Library, Yale University.)

derson's Orchestra and then tour with Teddy Wilson, staged the publicity session at Brunswick's New York recording studios with Duke Ellington and Chick Webb to promote the launch of a new record label. The photographer is Charles Peterson but the presence of another photographer in the image suggests this is a notable event. The event also parallels the historical record in that jazz spectators were interracial and of both genders. The three musicians appear to be comfortable: Shaw smiles, perhaps because he is publicly paired with his friend Chick Webb and the Duke, whom black and white musicians respected as the foremost composer of his age. The white members of the audience are all smiles. But all is not perfect in this photograph. Some of the black audience members appear to be a little nervous about being in such mixed company. The black man to the right of Oakley almost appears embarrassed to be pictured so close to a white woman, an image that might have produced a lynching had it occurred in the American South.

Figure 4.2: Jam session with Chick Webb (drums), Artie Shaw (clarinet), and Duke Ellington (piano), Brunswick Studio, March 14, 1937; Helen Oakley at Shaw's left. Photograph by Charles Peterson. Used with permission by Don Peterson.

An image that may have done more than any other to break down musical segregation was released by Hollywood's culture factory in January 1938. If there had been cognitive dissonance between knowledge about integration and beliefs about segregation, the Busby Berkeley musical *Hollywood Hotel* would have resolved this tension. Benny Goodman, Teddy Wilson, Gene Kruppa, and Lionel Hampton are all dressed in a similar manner and smiling as if integrated performances were normal. The image may have been made tolerable to whites because of motifs that maintained a racial hierarchy. Goodman and Kruppa are nearest the lights and the audience; they are biggest and brightest. Wilson and Hampton are connected by the tentacles of the shadow of an exotic-looking plant that subtly ties them to tropical spaces that were coded as black. Additionally, Goodman's and Kruppa's initials on the bass drum mark the territory as theirs and not that of Wilson and Hampton. It is a scene that may have comforted those who suffered from the cognitive dissonance of musical integration without permanent visual representation and educated those who had not yet experienced the dissonance. Nine years before Jackie Robinson appeared before hundreds of thousands of baseball fans, *Hollywood Hotel* may have been seen by an even greater number of moviegoers.[46] White and black bodies had been integrated.

The Presence of Voice

Advances in desegregation in music went largely unnoticed. These did not provoke protests or significant commentary. Indeed, the archive lists few written records of musical and bodily integration in jazz during the 1930s, though it became increasingly common. Vocal integration was, however, less easily accomplished. The voice lies closer to identity than even the body. As with the case of Abraham Lincoln, whose imagined deep voice reflected white beliefs about the sound of manly American-ness, assumptions about sound reveal the importance of the voice to the construction of identities.[47] Though race is predicated on skin color, race does not directly correlate with the body as evidenced by long-held concerns about passing and racial mixture.[48] We have a body. We never say that we are a body or that our body does something (unless it's something we don't want our body to do). Yet, we simultaneously voice concerns ("I voiced my opinion"—the voice asserts agency), have a voice ("I have a voice"—it is something that can be possessed), and are a voice ("I am a voice"—in other words, the voice is the whole person).[49] Thus, audiences could disembody instrumentalists from their selves but could not distinguish their voice from their selves, making visual integration easier to accomplish than vocal integration. The body can be dissociated from the self. We can pass as another race or ethnicity or even gender. But as Father Ong describes, we have a propensity to believe that the voice is the truest and surest marker of the self; in the spoken word, there is presence and power.[50] We believe, for example, that we can hear race when we cannot. And sometimes when we are confused about the race of the speaker, we try to make sense of it, as if voices carry that kind of meaning. Race, after all, is the product of only one of our five senses, and that sense can be easily deceived. We are tricked by the illusionist and entertained by the ventriloquist.

Racial assumptions of the interwar period emerge from songs performed by female vocalists in the 1930s that illustrate what black and white voices were supposed to sound like. In November 1933, Holiday joined an interracial group that Goodman had put together for a recording session. Jazz scholars are scathing in their assessment of Holiday's performance. The singer had gained fans for the same reasons that she remains popular today: her musical talent and her distinct phrasing. But the song recorded with Goodman, "My Mother's Son-in-Law," sounds little like the Holiday of legend. Critics who have been taken in by her deceptive autobiography write that she was nervous during the recording session, and so instead of her usual syncopated phrasing, her voice stays "straight." However, the myth that she was nervous was an invention, enabling the former prostitute to become innocent again as a first-timer in a recording studio with white and black men and a "big old microphone."[51] What has been missed is that Holiday likely *had* to sing "straight" because of the sonic racial expectations of the era. The record companies marketed Holiday as "black" or "white" depending on how she phrased her vocals and who the orchestra leader was; they and bandleaders dictated how musicians sounded, thus denying creativity of the kind that Holiday was famous for. On this recording, Holiday sounds inferior to popular, white vocalists of the 1930s; she most likely was directed to record that way by Goodman and studio executives, who may have worried that a "black" voice on a "white" recording would undermine their efforts to market race.[52] When Holiday recorded for "black" jazz groups, she returned to her usual style of syncopation and offbeat phrasing. Though Holiday recorded her syncopated style for "Miss Brown to You" and "What a Little Moonlight Can Do" (1935) with Benny Goodman and "Did I Remember" (1936) with Artie Shaw, the songs were marketed as two "black" acts, Teddy Wilson and His Orchestra and Billie Holiday and Her Orchestra, respectively. The invisibility produced by radio and phonograph players as usual masked integration.

A second example of the straight voice can be found in Martha Tilton's version of "Bei Mir Bist du Schoen," recorded with Goodman at his Carnegie Hall concert in 1938. The recording has the sounds of her appearing from offstage, a situation made necessary by the racially integrated performance. Jazz singers in the 1930s could sit in the bandstand when they were not performing. But at a concert featuring integrated musicians, the white woman could not have appeared onstage until the black men had left it—the "white" voice could not be integrated with black musicians. Though the radio audience knew that the instrumental part of the concert was integrated—the concert was famous—they would have known all of "Bei Mir Bist du Schoen's" performers were white because Tilton had not been onstage during the performance of the songs during the first half of the concert. How would they have known? Because white women did not play with black men even at an integrated concert and because a loud applause greets Tilton as she steps onto the stage. Not surprising, on this tune, Tilton does not vary from the beat, an example of straight vocals. The integration of voices was not yet acceptable.[53]

In 1938, Shaw and Holiday attempted to overcome the last bastion of racism in jazz, the segregation of the voice.[54] Sadly, they failed. Vocal integration was still intolerable to recording and broadcasting corporations. Holiday and Shaw describe a litany of racial incidents during their 1938 tour. Among other issues, she was sometimes prevented from joining the band on stage and occasionally got into fights with racists in the audience and after performances. In 1938, the Old Gold radio program forced Shaw to keep Holiday off the air during its coast-to-coast evening broadcasts because of Holiday's interpretations of the songs, or in other words her "black" phrasing. The cigarette company demanded that Holiday sing straight. Ultimately, she was forced out of Shaw's orchestra, but she had enough time to train, at Shaw's request, Helen Forrest, a white singer. Forrest learned Holiday's style in time to record "Deep Purple" (1939), in which the young, white singer imperfectly mimics the future legend. Though corporations had prevented him from performing with a black vocalist, Shaw was determined to have his white vocalist sound "black." While Shaw was trying to break the color line, Goodman remained attached to vocal segregation. In 1939, he performed again with Holiday for a radio broadcast under the Goodman banner. As one fan notes of the air check for "I Cried for You," Holiday sounds "as though she's not allowed to phrase the song properly." In other words, she sounds "white" again.[55]

That same year, Peterson put together a photo shoot for "Life Goes to a Party," a series that *Life* magazine ran on a consistent basis. In the photos, Holiday is joined by jazz artists and record executives. She sits or stands at the center of many the images, surrounded by men of three different races:

black, white, and Asian. Though photographs of integrated musicians and audiences had been published, *Life* mothballed the Holiday photo shoot.[56] Photographs of the integration of voices could still not be published. Unsurprising, Holiday largely gave up on vocal integration and turned to performing with an all-black group at the Café Society in late 1939.

Holiday's voice threatened the keepers of the official language. Speech teachers crusaded to maintain "proper" speech (i.e., the white, northeastern dialect) during the 1920s and 1930s. This kind of speech became the official language through national radio broadcasts and later national television broadcasts. One synecdoche for black dialect became what white teachers called stressing the wrong beat. In 1934, Emma Grant Meader warned, "[T]he failure to allow for the silent beat will syncopate the rhythm and in many cases spoil the meaning by stressing the wrong word."[57] Of course, syncopation is foundational to jazz and particularly for singers like Holiday who took the lead in stressing the "wrong" word. During the 1920s and 1930s, white elites worried that stressing the wrong beat was dangerous to the white soul. Thus, Anne Shaw Faulkner worried in 1921 that jazz put the sin in syncopation.[58] The language teacher Sophie A. Pray defined the purpose of speech education in 1923: "We avoid provincialisms, local dialects. . . . Language indicates social standing. We teach the form which gives the impression that the speaker belongs to the cultured class. We avoid vulgarisms, low class dialects."[59] The memoirs of the white saxophonist Mezz Mezzrow illuminate this pedagogy. He claims that he became a professional musician because he ran away from a sister who "corrected" his transcriptions of black blues lyrics.[60] Northeastern whites spoke the official language through which they enjoyed cultural capital.[61] The differences in dialect and intonation were critically important to the project of aurally distinguishing races and in doing so creating a vocalic hierarchy. Moreover, distinguishing between blacks and whites on the basis of voice became the last bastion of segregation in the invisible media of phonograph players and radio because the voice did not merely represent race. It was race.

The interwar voice carried meaning that was more proximate to truth than other forms of communication. The voice projects the self; bodily gestures do not. The voice is the *pneuma*, which means both "breath" and "spirit" or "soul" in a manner that gives life to communication. The voice moves; the voice occupies space. Even when captured, the voice is never static.[62] The voice, then, marks one's identity better than the body or any other signal the individual might produce through image, sound, smell, taste, feel, words, or otherwise.[63] It is the soul emerging from within and penetrating into the insides of others. This is not a new idea. Andrew Thomas Weaver noted in 1924, "Many teachers of voice have held that the voice is an indissoluble part of a mystical, occult, transcendental entity usually designated as *personality* or *soul*."[64] Where the body operates synecdochically to represent the person, the voice *is* the person. Walter J. Ong calls this "is" presence. He describes the relationship between words and sounds, arguing that language derives its power from the sonic, not the written. For Ong, the voice is the closest sensory coefficient of thought, that most human of all traits.[65] Because the voice is immediate and because it is fleeting, it is prototypically human.[66] During an era when a person's "color" was a key concern, it was, ironically, the voice more than the body that defined race.

Ong's description of presence is powerfully illustrated by a recording made in 1935. During the 1930s, academics and folklorists responded to the threat that old age posed to those few African Americans who remembered the years before emancipation. Researchers employed by the Library of Congress and the Works Progress Administration traveled throughout the south to interview former slaves. Most of these interviews are phonograph recordings of conversations held between white interviewers and black interviewees. Though tainted by interviewers' assumptions and by seven decades of accumulating nostalgia, these remain an extraordinary archive for researchers who study black culture in America. Fortunately, this archive is also helpful to those who study sound. A few of the former slaves were recorded by researchers like John Lomax (who also discovered Bukka White) and members of the American Dialect Society (ADS). In 1935, two academic members of the ADS traveled to Dunnsville, Virginia, where they found "Aunt Phoebe," a former slave born around 1850. There can be little question from the sound that Phoebe Boyd is old; she has a gravelly and weak voice, and her disjointed conversation abruptly shifts from the distant past to the immediate present, from growing tobacco to attending church to marriages, deaths, and relationships. Judging by Boyd's confusion of time during her interview, she may be suffering from advancing dementia.[67]

Five voices appear over the course of the eight sides that were recorded that day. Two of these are white male scholars, one is a female facilitator who seems to know Boyd, and another is a woman who has had prior experience recording the voices of former slaves. The device the speakers use is a portable, nonelectric phonograph recorder that was not much different from the machine Edison had invented in 1877. Understandably, the sound reproduced by the phonograph is rough. We return again to hearing pops, cracks, and fuzz, though these eight sides also create unnerving sounds.

One of these is a repeated back-and-forth swish that reminds us of a rocking washing machine. A second sound that the recording produces is an echo that occurs about halfway through each of the sides, a sonic effect likely produced by the manner in which the records were stored or by an original and near-identical warping defect in all the disks. Vocally, the recording provides ill-fitting truths. Boyd's voice may be an accurate guide to her soul, but it is not to events of the past. She is deferential to the white interviewers, and she is nostalgic about her days as a slave, though she often feels threatened—whether by the interviewers, the machine, or the memories—a feeling that she meliorates by raising her voice or singing in a manner that was probably taught to her in church. Both devices, the raised voice and the singing, remind us of the power of the sonorous envelope to comfort.

The recordings climax on the seventh side. At about the halfway point on this side, the recording decays into the swishy-washing machine effect as Boyd recounts a recent lynching. Though she claims that God is looking out for her, her claim is belied by the fear we hear in the pitch and speed of her voice and by her return to another lynching that she personally witnessed sometime over the course of her long life. As she begins to describe the scene, her voice is loud but falls off quickly into shame or sadness or insecurity. The end of her first refrain sounds very much like the end of a refrain of a lament spoken at a black church and is very nearly song-like. "And this man standin' there" is almost sung, ending with a downward glissando. The next horrifying phrase, "rope aroun' his neck," is compacted and glides downward again, Boyd's voice descending the musical scale. "And, uh" marks a period of confusion; Boyd is trying to make sense of her memories. "I tell you there's somethin' for you to look at" is very loud, rushed, and monotone as if Boyd is just trying to get through the retelling while comforting herself with the sound of her own voice. When she ends the series, "I don't know what I ain't seen in my life," every word is repeated in the same monotone except "what," which gets an upglide and exclamation point. Boyd does not know *what*; the scene cannot be translated into words for words cannot portray the awfulness of the event. But the sound of Boyd's voice translates the tragedy. That voice marks the lynching's presence.

Soon after, Boyd repeats the refrain "believe your eyes" as if to end the memory, then falls into silence only to be prodded by one of the interviewers to share whether or not the victim's family was present. Boyd shouts now, "Mother, father, brother, and all right there in the line." Incredibly, the echo found in all the eight sides emerges from the disk at this point, sonically doubling the number of witnesses to the lynching. And for the first time in the series of recordings, a pre-echo emerges, now tripling the witnesses, magnifying Boyd's voice and reminding us that time is out of sync. We do not know when the lynching occurred. But we can be certain by the tone of Boyd's voice—by her presence in 1850, 1935, and today—that she saw it. The recording, with its vocal tensions and its sonic distortions, has the power to make listeners nauseous. It is a black sound, but not because Boyd or the victims of the lynch mobs are black—it is a black sound because of how the voice betrays the subject matter. In the 1930s, speech teachers urged students to speak in a bright tone. Boyd speaks in the darkest of tones, coloring our perception of race. But only if we listen.

During the interwar period, sonic assumptions hosted a belief in black sound that was closely related to blues. The term *black music* was used at least as early as 1924.[68] That the phrase could have then meant a variety of things—foreboding music, African rhythms, or the blues—is a testament to the confusion of sound and vision, the psychological term for which is synesthesia. The conflation of colors and sounds has an ancient heritage extending back to the Romans, whose term *colores* meant both color and style. Politicians like Cicero employed the term to refer to the tropes and figures that were commonly used to make persuasive arguments.[69] Seneca the Elder, Cicero's contemporary, describes how a speaker's tone might color listeners' interpretations of the argument. That tone could cast light or darkness on events, facts, and personalities. Thus, the intent of arguing was the winning of at least "*colour*able" support.[70] While this seems like a wide variety of meanings for a single word, *colores*, it is not. *Colores* are the rhetorical ornaments that make an argument subtly persuasive and influences the minds of auditors. Henry David Thoreau made sense of *colores* in a colorful simile: "As the last drop of wine tinges the whole goblet, so the last particle of truth colors our whole life. It is never isolated, or simply added as treasure to our stock. When any real progress is made, we unlearn and learn anew what we thought we knew before."[71] For Thoreau, the word *color* still has its triple meaning as a noun describing a visual quality (henceforth *color*), as a term for rhetorical figures (henceforth *colores*), and as a verb connoting to change, to influence, or to distort (henceforth *colour*). The *color* of sound cannot be seen. But the *colour* of sound can be heard.

Being *coloured* by the voice of a speaker was a particularly troubling concern during the interwar period when *color* referred to race and when speakers were often invisible. Voices were supposed to be paired with the color of bodies. Bodiless voices threatened identity categories that were predicated on visual racial distinctions. In *The Psychology of Radio,* A. H. Cantril and

G. W. Allport worry, "Voices have a way of provoking curiosity, of arousing a train of imagination that will conjure up a substantial and congruent personality to support and harmonize with the disembodied voice."[72] Radio, phonographs, and telephones compounded the problem of colouring during the interwar period because these prevented audiences from physically seeing the color of the speaker.

The term *colores* has also long been applied to music, at first to "ornaments of style" and later to colors more broadly.[73] Public-speaking scholars adopted this musical understanding and applied *colores* to the sound of speech and not just to tropes and figures. Everyone's voice, according to early twentieth-century scholars, had a color, and that voice could be modified to take on another, though similar or appropriate, color.[74] It wasn't hard for auditors to extend the sonic concept back to race. Funk & Wagnalls *Comprehensive Standard Dictionary of the English Language* (1933) illustrates the conflation in its definition of *color*: "I. *t.* 1. To give a color to; dye; paint; tint; stain. 2. To misrepresent; modify; colorable (a. coloration, n. colored); give a tone to; II To change color; blush . . . Having color; of a dark-skinned race; embellished or exaggerated."[75] To misrepresent, to give a tone to, and of a dark-skinned race suggests a conflation of many ideas, the nexus of which lie in this one word.

But it was not just red wine in the goblet, there were white wines, too. The trouble was that the colors were so subtle as to demand an outside source—the appearance of the body—to make sense of them. Sound media created racial confusion during the interwar period by altering Americans' communicative practices. The Kodak Brownie camera and movie theaters had become popular before World War I, and these usually transmitted the race of the individual whose image was captured. After the war, Americans of every background had access to radios, telephones, and phonographs. By World War II, almost 90 percent of American families owned a radio; 40 percent owned a telephone; phonograph ownership was in between.[76] Because of the ubiquity of these technologies, sound presented the potential for challenging racial categories in the 1920s and 1930s and enables twenty-first-century scholars to listen to constructions of race during the period.[77] We can now see and hear *colour*.

Holiday's "Strange Fruit" exemplifies the rhetorical concept *colores*. Of course, music has long been influenced by rhetoric and, in particular, the idea that music should persuade. Baroque composers, for example, hoped to effect and affect listeners by using rhetorical techniques like repetition, imitation, and dissonance.[78] Not surprising then, that protesters have a long history of pairing words with songs in the belief that their tunes would influence listeners. The best protest songs not only combine words with music but also evidence a manipulation of sound in a manner that supports or deepens the words. Louis Armstrong's "What Did I Do to Be So Black and Blue" preceded Holiday's "Strange Fruit" by ten years, but it is because the latter exerts such tremendous rhetorical force that it is remembered as America's first great protest song. Originally written as a poem by a left-wing teacher in New York City, the lyrics to "Strange Fruit" exemplify the rhetorical trope of metaphor. Abel Meeropol had no interest in real fruit hanging from trees—he was describing the hanging of black men. The poem is not only maudlin but also disgusting ("bulging eyes," "burning flesh," "sun to rot") and yet won the favor of New York leftists. Meeropol set the poem to music, and though he continues to receive credit for the music, the vocal delivery is entirely Holiday's.[79]

Before recording the song, Holiday honed it to her satisfaction while singing to integrated audiences at Café Society. After her painful stint in Shaw's big band, the torch singer found a place where the patrons saw race less vividly than probably anywhere else in the United States. The club played host to Ellison, Charlie Chaplin, Richard Wright, Duke Ellington, Goodman, and many other white and black literary, film, and musical artists.[80] Café Society was also an explicitly political venue where left-wing views were encouraged, and, thus, Holiday sang in a safe place from which she could work through the sonic kinks in "Strange Fruit." By the time she recorded the song, it had become so powerful that Holiday was closing her set with it, without the possibility of encore.

The 1939 recording features Holiday employing a wide array of vocal and rhetorical tricks to emphasize words and provoke emotional reactions.[81] The south's purported goodness, for example, gets an ironic treatment when Holiday twists phrases like "sweet and fresh" while eliding "gallant" into something sonically less than the full word. She uses a kind of reverse onomatopoeia when she infuses words with sounds that reiterate the meaning of the word. Her intonation of "sudden," for instance, is rapid, thus turning the word into an example of itself. When she forces out the word *bulging*, she imitates with her voice the visual appearance of something being forced outward. The word *breeze* is elongated, and the letter *b* in *blood* drips from Holiday's lips like the life force of the victims she describes. When Holiday sings *drop*, her voice briefly ascends then descends in a long glissando. At the end of the dragged out *drop*, Holiday's vibrato sonically mimics the tension of the long rope bouncing at first then quivering, then remaining still. Her voice has gained in intensity until this moment but then fades out, suggesting that it is at this point in the song when the lynching has occurred and when life is ending. That point is hammered home by the moment of silence in the recording immediately after the lynching and by the anticlimactic final sentence, "Here is a strange and bitter crop." Throughout, Holiday uses syncopation, dissonance, and uneven vocal projection to create an uncomfortable atmosphere that mirrors the lyrics. These are just some of the *colores* that she uses in "Strange Fruit."

Like FDR, Holiday understood how to hold an audience's attention by rendering words into sonic persuasion. But Holiday must have been concerned with her audience's likely reaction to a gruesome tune about lynching in America, a problem exacerbated by a lengthy introduction through a somber trumpet and piano colloquy that the producer added to make its length appropriate to the audience's expectation of three-minute recordings. Holiday's opening refrain is painful to listen to both because of the words and because of the passion the artist brings to the sound:

> Southern trees bear a strange fruit.
> Blood on the leaves, and blood at the root.

The injection of the word *and*, a word that did not have to be in the poem or the song, suggests that Holiday was worried that she might lose her audience because her use of "blood on the leaves" would have defied so many expectations. Holiday was famous for love songs like "Miss Brown to You" and "I Must Have That Man." But if she defeated her audience's assumptions by singing about blood, how could she keep them listening? By intoning *and* in a manner that made this rather common word stand out among its neighbors. Like FDR's "is" in "the only thing we have to fear is fear itself," Holiday's forceful conjunction begs the question, "And what?" The word then blends into the next phrase, "blood at the root," dragging the audience along with the voice. They have been held in place by the exaggerated *and* and by its transition into the remainder of the poem. By holding her listeners, Holiday *coloured* their perception, winning assent to her argument that black men should never be lynched. There is something beyond a recorded voice in "Strange Fruit." There is presence.

The effect of Holiday's phrasing, like the voice of Boyd, is uncanny. Sigmund Freud wrote in 1919, "The sense of the uncanny attaches directly . . . to the idea of being robbed of one's eyes." This problem is summarized in the German term *heimlich*, which means simultaneously that which is familiar and comfortable and that which is concealed or kept hidden.[82] And while scholars have argued that all the sounds that emanated from telephones, radios, and phonographs were uncanny, I argue that some sounds were (and remain) particularly so.[83] Following Freud, Mladen Dolar argues about recording devices, "There is an uncanniness in the gap which enables a machine, by purely mechanical means, to produce something so uniquely human as voice and speech. It is as if the effect could emancipate itself from its mechanical origin, and start functioning as a surplus—indeed, as a ghost in the machine; as if there were an effect without a proper cause."[84] For Dolar, the ghost in the machine creates powerful effects, a surplus that in Holiday's recording makes her argument particularly persuasive.

The uncanny erases "boundaries between interior/exterior, self/other, and subject/object."[85] From Ong, we understand that voice makes the interior exterior, pushing out the presence of the mind/soul from the center of the body into acoustic space and making individuals present. The words spoken by Boyd and sung by Holiday are made powerful because they are uncanny. The voices remain, even today, uncanny because they emerge from a past to tell white Americans about the horrors their ancestors caused and black Americans about the horrors their ancestors endured. The voices also emerge from the ether, exemplified by the pre- and postecho in the recording of Boyd, leaving us with nothing but presence. And finally, the voices are uncanny because these trouble our ability to identify according to visual categories like black and white. Boyd and Holiday *colour* our perception of the experience of race (rather than race itself) through black sounds.

Making the Voice Visible

The critic Rudolph Lothar, writing about phonograph machines in 1924, complains, "The machine demands that we give bodies to the sounds emanating from it."[86] Likewise, in a 1923 essay, a reviewer reasons that some people "cannot bear to hear a remarkably life-like human voice issuing from a box. They desire the physical presence. For want of it, the gramophone distresses them."[87] Until the advent of sonic technologies, humans had relied upon visual cues to pair the voice with the body. Getting lost in radios, phonographs, and telephones, listeners became disturbed by selfhood without visual clues.[88] In the 1890s, whites told jokes about African Americans who were fooled by noises emanating from gramophones. And though ostensibly told because the whites supposedly knew better, the jokes illustrate that whites were also having trouble identifying the speakers.[89] Whites displaced their anxieties about invisibility and identity onto blacks. Without visual cues, white listeners of minstrel-show recordings in 1890s, for example, may have had trouble recognizing the race of performers.[90] That anxiety would in turn explain the popularity of recordings that vividly portrayed the ethnicity of speakers by exaggerating accents.

Serial radio broadcasts of the interwar era exploited these anxieties. *The Shadow* radio show introduced the first superhero to the airwaves in 1937. Orson Welles provided the protagonist's voice with an effect that makes it sound as if he is on a telephone. Though not entirely invisible (after all he was named the Shadow), one enemy said of him, "You can hear his voice, but you can't see him." On a radio show, it could have been no other way.[91] During a 1937 episode, "Deathhouse Rescue," the superhero's girlfriend, Margot Lane, confuses a victim who will soon be helped by the Shadow. The victim asks Lane about the Shadow's true self. Lane responds, "It's hard to tell whether I really know the man, or only his shadow."[92] While the dictionary defines *uncanny* as that which suggests superhuman powers, Freud argues the uncanny is also that which is familiar yet alien. The Shadow, according to the person who knew him best, fit all these definitions. The uncanniness of the Shadow reveals itself in another racist joke that illustrates insecurities about identifying invisible voices. A joke dating to the 1940s describes the death of Lamont Cranston, the fictional alter ego and voice of the Shadow on the radio program. According to the joke, the producers of the show could not find a suitable replacement other than a black janitor whose masculine voice was perfect for the role. During his first on air appearance, however, the janitor began, "Who knows what evil lurks in the hearts of men? The shadow do." The joke reminds us of the belief that sound technologies would reveal the truth and that the voice would always betray its nature. But the existence of the joke suggests people were concerned that it might not.[93] Thus only poor grammar tripped up the black Shadow.

The joke also reminds us of the most famous of the early serials, so famous that it compelled consumers to purchase radios. First broadcast in 1928, *Amos 'n' Andy* featured white writers and actors Freeman Gosden and Charles Correll, who portrayed two African American ne'er-do-wells to an audience that quickly grew to forty million listeners, almost of all of whom were white. Borrowing from blackface minstrelsy and a theatrical tradition that extended back to early nineteenth-century white actors who played a purportedly black character named Sambo, Gosden and Correll turned the denigration of African Americans into an incredibly profitable enterprise. Most scholars who have studied the show argue that Gosden and Correll promoted racial divisions and stereotypes.[94] But scholars have failed to grasp the power of radio: it was not visible. As we now know, voices are constructed as black or white. And during the formative period of American radio broadcasting, the visual classification system of race had to be transformed into an aural classification system. Whites, for example, couldn't be *seen* to have invited blacks into their living rooms via the radio, yet they could invite actors who sonically passed as black. Though the minstrel shows that featured white performers in blackface had become passé by 1930, whites and blacks were familiar with the cross-race conventions of the genre. Susan J. Douglas argues that *Amos 'n' Andy* actually helped to blur distinctions between whites and blacks exactly because the actors were invisible: "White listeners weren't simply laughing at black folks; they were also laughing at an only slightly exaggerated version of themselves. All too many white listeners, although most would never admit it, identified with *Amos 'n' Andy*."[95] During an age of rapid modernization, industrialization, automation, and unemployment, the show was really about shared anxieties. Not only were whites able to identify with black characters but the program also occasionally dropped the charade of blackface, revealing the actors to be thespians rather than African Americans who happened to wade through the radio waves into the homes of tens of millions of white Americans.

Until 1939, the toothpaste company Pepsodent sponsored the national broadcast of *Amos 'n' Andy*, and so most shows began with a brief tune played on organ with a voiceover encouraging listeners to buy toothpaste. During the August 19, 1936, broadcast, "Eighth Year on NBC," Amos and Andy celebrated with a special guest, the movie star Walter Huston, who had played the straight-talking Abraham Lincoln in D. W. Griffith's 1930 film, *Lincoln*.[96] The radio play's plot is simple. Amos and Andy are visiting Hollywood and borrow a car that runs out of gas in the hills northeast of Los Angeles, so the two find a house where they hope to use a telephone. The house belongs to Huston, who has no telephone but offers his assistance in the form of gasoline and food at no charge. Over the next few minutes, the three men become fast friends, not a likely result had Amos and Andy actually been black and lost in a lily-white neighborhood. In a deeply segregated, Jim Crow–following, post–separate-but-equal America, the idea of two black men being treated as equals in Walter Huston's living room could have been revolutionary. But it was not. Instead, it was sponsored by a product that whitened.

Sonically, the characters voiced by Gosden and Correll exhibited what many would have considered to be variations of a black dialect. The actors replace the *th* sound with *d*, drop *g*'s from the ends of words, and simplify verb tenses.[97] Correll's deep bass voice also signified blackness and, in particular, the hypermasculinity of those black men who threatened to disturb sexual norms.⁹⁸ Yet, their voices remain remarkably accessible; sonically, they are not particularly "other." Huston speaks in the "white," official dialect. Finally, the announcer who hawks Pepsodent at the beginning of the program and returns to formally introduce Huston at the end of the program also speaks with a "white," northeastern accent. This is a white production for a white audience. Notably, in the outro, Correll makes fun of the invisibility of skin color on radio in response to praise from Huston: "I was blushin.' Ain't nobody heard me blushin' on the radio." Like the Shadow, the actor draws attention to the invisibility created by radio and the difficulty of connecting presence to the body. In doing so, *Amos 'n' Andy* naturalized and made understandable the voices emanating from millions of radios. Noises (including voices and dialects) contain the other, and if these can be purchased and owned, the other can be domesticated and even incorporated into one's own identity and selfhood.⁹⁹ *Amos 'n' Andy* constructed blackness through sound even as the consumption of the program deconstructed blackness.

The Invisible Scarlet O'Neil, the first female superhero, further illustrates the easing of the uncanny by training listeners to become comfortable with invisibility. In this sense, she exemplifies the desires of Adorno and Arnheim, who believed that music should be heard without eyes. In one episode, O'Neil uses her invisibility to trick a band of gangsters including a tough named Eddie who are holed up in a nightclub's back room. We pick up the action with the invisible O'Neil in the gangster's room:

> Scarlet said to herself, "They'll be making an announcement, Eddie—one you're not expecting to hear!" She walked over to the radio and turned down the volume. One of the men said, "Here it comes, boss," and they all laid down their cards and turned toward the radio to listen.
> It was Scarlet's voice they heard. She said, each word ringing clearly, "Ladies and gentlemen, we interrupt this program to bring you a special announcement. The execution of Everett Hadly did not take place tonight. Last minute evidence revealed that he was framed. The police are on their way to capture the real killers who are said to be hiding in the back room of a west-side night club."
> Scarlet was watching their faces as she spoke. She saw the smiles fading into looks of astonishment and unbelief and then into stark terror.¹⁰⁰

O'Neil mimicked the invisibility produced by the radio and mimicked again the sound of the radio with her special announcement. That the radio could be employed to deceive was not new when this adventure was first published. Welles had famously deceived millions of radio listeners with "The

War of the Worlds" in 1938. But in the O'Neil adventure, the invisibility of the speaker did not deceive. Only the thuggish and fictional gangsters are duped and made to feel the terror posed by the uncanny voice in their room. Consumers of the comic knew their hero was in the room, and they slipped into her groove, thus performing her invisibility themselves. They were no longer made to feel awkward by the uncanny nature of the phonograph and the radio. Instead, they adapted to its invisibility.

By the disappearance of O'Neil, the uncanny nature of sonic machines is itself beginning to disappear. The sound of voices emanating from radios, telephones, and phonographs has become familiar. When O'Neil breaks into a radio program to begin her deception, she says, "We interrupt this program," a phrase that had made its way into the sonic vernacular. And so after 1941, the uncanny is eased because sonic machines no longer produce concerns about invisibility. Before invisibility was normalized, however, it created opportunities for challenging visual constructions and a window through which twenty-first-century scholars can study the race and color of sound.

Superheroes and jazz music were particularly well supported by America's youth, and it was young people who led the way to integration. Indeed, try as they might, media corporations were not able to prevent vocal desegregation. Thanks to the invisibility produced by radio and phonograph players and to the bodily integration of instrumental jazz music, interpretations of vocals could run counter to expectations. Already in the 1920s, thousands of white youths were seeking out "black" music. Fitzgerald wrote in 1923 of young and wealthy white students studying and imitating black dialects to the horror of their parents.[101] Words that white kids had learned like "hep-cat," "viper," and "daddy-o" entered the mainstream vocabulary, and so, too, had the phrasing used by black singers like Holiday. Young people often resist adult norms through the adoption of a manner of speaking and musical tastes that are antithetical to the official language and official culture.[102] Because radio lost much of its uncanny power to disturb, young people by the 1940s were able to reject the segregation of voices. It is to them that credit must go for both physical and vocal integration.

5

Sounds of War

The ancient Greeks told tales of the sirens whose voices so enchanted seamen that the mariners lost their minds and crashed their ships onto the shores. The sirens' calls were irresistible. Even Odysseus, after plugging his mariners' ears, had to be tied to the mast to prevent himself from losing his mind, his ship, and his life. In history, it is one of the most powerful of sounds, though mythological, and demonstrates once again sonic persuasion. Some sounds can make people commit suicide.

In 1937, the American poet Langston Hughes employed this motif during a visit to Spain, a country enduring a national suicide in the form of a civil war. Hughes enjoyed visiting a Madrid bar where performers sang, played, and danced flamenco. He fell into rapture every time he heard one singer in particular. For the African American expatriate, the passionate and rhythmic style of flamenco, combined with the singer's voice, reminded him of the music of his home:

> The guitars played behind her, but you forgot the guitars and heard only her voice rising hard and harsh, wild, lonely and bitter sweet from the bare stage of the theatre with the unshaded house lights on full. This plain old woman could make the hair rise on your head, could do to your insides what the moan of the air-raid siren did, could rip your soul-case with her voice. I went to hear La Nina many times. I found the strange, high, wild crying of her flamenco in some ways much like the primitive Negro blues of the deep South. The words and music were filled with heartbreak, yet vibrant with resistance to defeat, and hard with the will to savor life in spite of its vicissitudes.[1]

Hughes recognizes a sound that simultaneously fills him with heartbreak and yet wills him to savor life and resist defeat. The sonic analogy to the air-raid siren is doubly apt. Like the singer, air-raid sirens climb in a crescendo and then rise and fall. From the 1930s until the 1960s, their moaning, for that is truly how they sound, penetrated sonic envelopes and encouraged audiences to respond with fears of bombing raids and death. But the purpose of these sirens was not to terrorize listeners into paralysis or hysteria—it was intended to save them. The heartbreak would come with news of the loss of friends and relatives, but listeners would survive, would resist defeat, and would will themselves to live. The singer who so enraptured Hughes was indeed a siren who lured him toward tragedy, forewarned of death, and yet encouraged his will to live.

Hughes was an avid jazz fan and may have recognized in the siren's flamenco call his beloved blues chords and, in particular, the commonly employed minor third. That chord is, not surprising, the chord employed in air-raid sirens. Like the blues, air-raid sirens moan in a rising and falling pattern, with long glissandos bridging lengthy passages of two minor thirds. The sound of those minor thirds is "down" or, synesthetically speaking, blue. These are not the sounds of triumph or joy but of struggle and loss. And yet, like the blues, the minor chord can be employed as a way of resisting struggle and loss. Indeed, the very act of turning tragedy into sound reminds us of how all humans are capable of elevating themselves above heartbreak and of turning sadness into beauty. Air-raid sirens may be frightening and may remind listeners of moaning, but they are not ugly. Far more offensive sounds are in the human repertoire.

The modern air-raid siren also enables identification with others and the creation of the self. And its wail coerces individuals to die stoically or perform citizenship. It is as close to iconic as any noise in the American soundscape. It is a sound that we hear at a suggestion, even the suggestion of an image, and one that we have been trained to listen for. Siren calls have persuaded individuals to take particular courses of action—empathizing with the British or seeking shelter and surviving, for example—prescribed by the government so that the nation could live to fight against its enemies. Though the government was aware that only a few people on the margins of atomic-blast zones might be spared by ducking and covering, bureaucrats in the 1950s employed air-raid sirens to frighten citizens into obedience and to buttress public opinion.

The meaning of the sound of bombs, on the other hand—from the falling sound of Wile E. Coyote in the Roadrunner cartoons, to Lyndon Baines Johnson's voice-over film footage of a nuclear explosion in the Daisy ad of 1964, to the absence of sound in Pentagon propaganda films of bombs dropped on Iraq in 1991—has been contested in political and popular culture. These sounds include whistles, voices, silences, and ticking and operate in a variety of persuasive registers, though most commonly through irony or fear-mongering. While some of the sounds attempted to defeat the sonic power of air-raid sirens, others exploited the sounds with effects as powerful as the siren call.

As always, I argue that sound can be read. Even noise persuades in some form or manner. Certain sounds related to war, like silence and the voice of God, are particularly persuasive.

Siren Calls

Hughes would have heard the sound of the air-raid siren again after he returned to New York City. These became a soundtrack for two violent periods in Western history, first, during World War II when the sirens reminded Americans of their country's roots in England when England's people were enduring conventional air raids and, second, during the cold war when the sirens warned of atomic attack. If Hughes had listened to CBS during the second Great War, he would have heard Edward R. Murrow reporting from London on the almost-nightly German bombing raids. Murrow's reports were so widely respected that they contributed to America's entry into the war on the side of the British.[2] The air-raid sirens recorded for those CBS broadcasts helped to accustom Americans to the war by permitting audiences to listen through the ears of their English counterparts.

Rhetorical scholars have analyzed rhetoric for tropes that enable auditors to identify with others. Kenneth Burke, in particular, gave a renewed emphasis to the rhetorical concept of manufacturing identification through words.[3] For Burke, rhetors are at their most persuasive when they identify with the properties of others in the form of possessions, emotions, citizenship, and status. The rhetor builds ethos by informing auditors that he or she is like them and shares their hopes, interests, and fears. In that common ground, rhetors and auditors identify with each other. So far, however, the work of rhetorical scholars on identification has been mostly through words, though visual scholars have also begun to work with image and identification. Not surprising, sound also creates common ground and identification, perhaps more powerfully even than other senses.

Murrow's reports were intended to create identification. One of the most famous reports was filed on August 24, 1940, during a live broadcast from Trafalgar Square as London experienced an air raid.[4] The report titled "London after Dark" opens with a few seconds of the distant sounds of multiple air-raid sirens, setting a foreboding tone for radio listeners who were tuned into the CBS network. Understanding the power of radio, Murrow concentrates his commentary on the sounds around him: "The noise that you hear at the moment is the sound of the air-raid sirens" for those few Americans who had not heard a recent report about the war in Europe. Building atmosphere, Murrow warns his audience, "I'll just let you listen to the traffic and the sound of the siren for a moment." He understood that Americans wanted to find common ground with the English-speaking Londoners who traveled by automobile or by foot to air-raid shelters. The fear that Americans vicariously experienced through their radio speakers was not intended to cause panic, paralysis, or hysteria. Rather, Murrow restrained his audience through his calm and steady reporting and diffused the drama by reminding Americans of the stoic Londoners, who willed themselves to resist defeat. In this manner, Americans imagined a community they shared with the English, braving the air raids together with stiff upper lips. Rather than attempting to connect through the silent medium of print, Americans learned to empathize with the British through the live medium of sound.

Murrow understood that the moan of those sirens might frighten listeners into inaction, a possibility he wanted to avoid. Instead, he hoped the American people would support the British war effort. But as Murrow also suspected from the German experience of radio (and from critiques of sound dating back to Plato and Immanuel Kant), sound broadcasts had the potential to provoke extreme emotions. An essay published in *Variety* in 1939 warns of radio, "While it does not create the tensions of the day, radio elongates the shadows of fear and frustration. We are scared by the mechanized columns of Hitler. We are twice-scared by the emotionalism of radio. Radio quickens the tempo of the alternating waves of confidence and defeatism which sweep the country and undermine judgment. Radio exposes nearly everybody in the country to a rapid, bewildering succession of emotional experiences."[5] For the editors of *Variety*, radio was all too live. But Murrow maintained a steady and calming presence on the radio, never sounding excited or frightened and always keeping an unsentimental air of unbiased reportage. Even so, Murrow was incredibly savvy about how he sounded and employed phrasing in a manner that reiterated his assertions and beliefs. Murrow used pauses to heighten listener interest and intonation to emphasize words like *edge, dead,* and *flat*.[6] Additionally, Murrow employed quotidian sounds to remind his listeners that even in the midst of war, life continued. The voices that can be heard "under" Murrow are normal, and the footsteps he records are universally pedestrian. Through the radio, Americans stood in the shoes of the English. Murrow very much wanted Americans to hear this lack of hysteria and panic: "One of the strangest sounds one can hear in London these days, or rather these dark nights, just the sound of footsteps walking along the street, like ghosts shod with steel shoes." Of course, these were ghosts that Americans heard through their radios—the bodies were invisible. Yet, the voices and the footsteps were real and, what is more, rhetorically forceful. The voice of Murrow successfully accustomed Americans to the sounds of war and helped to convince them of the necessity of joining England against Germany and Italy. Indeed, his vocal power transcended the war, elevating him into a voice that Americans would believe even after the curtain was torn away, as when Murrow appeared on the program *See It Now.* During World War II, Murrow used his voice to inform Americans about European events and to reassure that rather than fear the horrifying sounds of war, they could steel their resolve and squarely face the evils of Nazism.

The meaning of sound, as we can hear, is multifaceted and often operates at a subconscious level. Sound *c*arries the *p*ower of *p*resence that the *p*ermanency of *p*rint *c*annot *c*onvey (alliteration is entirely sonic, for example). The voices that emerged from the ether of the relatively new technology of radio only strengthened the power of the voice and so, too, the power of other sounds like the air-raid sirens that are present throughout Murrow's August 24, 1940, broadcast. The sound of those sirens, however, held the possibility of even-greater manipulation because they did not produce words that critical thinkers might analyze and debunk. Though communication scholars had devoted much attention to the power of sound in the 1930s, they had not yet developed a critical vocabulary for understanding the persuasive force of sound. Sadly, efforts to learn how to critique sound stalled at the beginning of the intensely visual cold war. With the mass introduction of television, scholars largely abandoned the study of radio and sound.

Even before the advent of television, the American government was deeply involved in learning how sound could manipulate the national psyche. Though print propaganda had been employed during World War I by the Committee for Public Information, efforts during World War II to manipulate public opinion through a variety of media became increasingly more academic, pervasive, and subtle. The Office of Civilian Defense began in 1940 under the leadership of New York Mayor Fiorello La Guardia. Against the wishes of his superiors, however, La Guardia ignored the importance of psychological manipulation to home defense and was eventually forced to appoint Eleanor Roosevelt as his assistant director. She and her husband understood the importance of psychology to promoting civilian morale and to the success of the war effort.[7] By 1944, the Office of War Information had assumed control over civilian morale. That office's work was typified by propaganda posters and cheerleading for America's efforts.[8] Though the office was closed later that year, its work continued under a series of emergency contracts with universities, contracts that continued until 1958. Those universities created or expanded communication, sociology, and psychology departments whose scholars applied themselves to determining how Americans would react to a variety of stimuli related at first to conventional war and, after 1949, to atomic war. The National Opinion Research Center (NORC) at the University of Chicago, for example, was tasked with studying "community disasters," which made earthquakes and tornadoes analogous to chemical weapons. Soon enough, NORC scholars would also study the psychological impact of atomic weapons.[9] The federal government funded institutes for communication research at a variety of universities to study media effects, such as, how audiences would interpret news about war and educational films that were designed to train Americans to react appropriately.[10] The largest of these programs was Project East River, a government-sponsored collaborative effort carried out by researchers at the universities of Cornell, Columbia, Harvard, Johns Hopkins, and Yale, MIT, the University of Pennsylvania, and the University of Rochester. Beginning in 1951, these scholars put together ten reports about how the American people might survive atomic warfare.[11] The affiliated universities came up with fifteen general recommendations, most notably that "the civil defense program must place first reliance on the efforts of the individual and the community to increase chances of survival, to minimize damage, and to recover as quickly as possible in the eventuality of an enemy attack."[12] The correlation between these recommendations and the meaning of the air-raid siren is not accidental.

The National Security Act of 1947 made explicit the design of these programs. That law aimed to determine how both foreign and domestic audiences could be manipulated into behaviors that would support the policy aims of the American government. On the home front, this would come to mean doing everything possible to survive an atomic attack so that the United States could successfully fight back against aggressors. By an executive order of 1950, President Harry S. Truman combined a variety of civil-defense operations efforts into the Federal Civil Defense Administration

(FCDA).¹³ That administration continued the sponsorship of academic studies and built on the work of previous propaganda efforts to sell civil defense to Americans.¹⁴ Those efforts were typified by the production of pamphlets, comic books, and characters like Bert the Turtle, who was made famous by the educational film *Duck and Cover*. The FCDA's message was further supported by patriotic media outlets that were themselves influenced by the propaganda. Dovetailing with the East River Project, for example, *Collier's* magazine published an article warning that the potential for defeat in a future World War III would not be the fault of atomic weapons but rather the fault of Americans' lack of preparation for survival. One last cold-war federal bureaucracy to enter into the battle for Americans' hearts and minds was the Psychological Strategy Board. This organization was tasked with foreign psychological operations and forbidden from participation in home-front activities yet became a behind-the-scenes player in adapting foreign psy-ops strategies for use against Americans.¹⁵

All of these efforts addressed a series of related conundrums. How could Americans be convinced to be afraid of atomic attack without being paralyzed by fear? Furthermore, how could Americans be convinced that they were vulnerable to attack and yet enjoyed the most powerful military force on the planet and that atomic warfare was deadly yet survivable? And how could government propaganda preach self-sufficiency during a period when Americans would most need assistance while maintaining popular support for the government's many bureaucracies?¹⁶ The propaganda produced by civilian-defense agencies was supposed to manage emotions, to train behavior, and to suppress terror, but, as we now know, many of these efforts failed to balance fear with hope.¹⁷ Though the pamphlets and films were designed to train schoolchildren and their parents to respond appropriately, we only need ask those who experienced this propaganda to know that it produced society-wide nightmares that ultimately manifested in 1950s B-movies like *Godzilla* and B-sides like Bill Haley's "Thirteen Women (and Only One Man in Town)."¹⁸

Duck and Cover (1951) is a horrific film. The propaganda vehicle opens with a catchy and happy tune sung by a group of four men and women in a style near that of barbershop quartets. The cartoon that accompanies the opening is Bert the Turtle, who hides inside his shell when he sees danger. After abruptly switching from cartoon to a documentary format and beginning about four minutes into the film, an air-raid siren moans in a series of three long stretches that carry for fifty seconds overtop images of children calmly seeking shelter. The first moan is disconcerting, but we are relieved at its end. However, it starts up again immediately afterwards to illustrate a second example of ducking and covering. Again we are relieved when the moan ends, and again we are disturbed when the siren wails for a third time. The effect is disconcerting and leads us to think that the moan will begin for a fourth time, an effect that might have left audiences in the 1950s to anticipate a never-ending series of air raids. The effect approximates what the sound theorist R. Murray Schafer described as keynotes that "are heard by a particular society continuously or frequently enough to form a background against which other sounds are perceived. . . . They act as conditioning agents in the perception of other sound signals." In this case, the sirens conditioned auditors to obey the commanding voice over in *Duck and Cover*.[19] As with the Londoners of Edward R. Murrow's broadcasts, listeners of *Duck and Cover* were trained that they would be subject not to one attack but to repeated attacks and were conditioned to believe that the siren's moan would become a frequently repeated signal of danger and destruction. When the training in the film was born out in everyday life through repeated air-raid drills, the testing of air-raid sirens on a regular basis, and the emergence of the sounds of the sirens in popular culture, Americans became conditioned to fear the moan against which they had to continue living. Rather than prompting Americans to behave appropriately and to survive, *Duck and Cover* produced fear and anxiety and may have contributed to uncritical support for hunting down pinkos and commies during the McCarthy era.

The sounds in the film are overwhelming. Michel Chion argues that to think image and sound automatically complement each other is to dismiss the power of sound to manipulate and persuade. Chion's theory of added value dictates that sound enriches images in both complementary and ironic ways and that, improperly employed, sound can destroy the meaning of the image and even the voice.[20] Applied to cartoons and to children, sound artists must be particularly careful to sync sound to image, a problem to which animators gradually adjusted.[21] The juxtaposition of cartoon with air-raid siren ironically undermined one purpose of the air-raid siren. The extreme contrast of a fun cartoon and the life-threatening moan of an air-raid siren defeats the calming purpose of *Duck and Cover*, not to mention defeating the purpose of air-raid sirens.[22] Rather than training children to coolly seek shelter when faced with the threat of an atomic attack, the film produces fear.

The media theorist Marshall McLuhan also worried over the effects of this kind of contrast. Writing about the hot media of film and the cold war, McLuhan begins with the example of a print advertisement for an insurance company "that featured Dad in an iron lung surrounded by a joyful family group [that] did more to strike terror into the reader than all the warning wisdom in the world." If print media could produce such a catastrophic reaction, then film media had to be very careful: "As for the cool war and the hot bomb scare, the cultural strategy that is desperately needed is humor and play. It is play that cools off the hot situations in actual life by miming them."[23] Given that cartoons are a form of "cool media," according to McLuhan, the cartoonist Chuck Jones got it right in terms of both the medium (cool cartoon to portray cold war) and the message (play to mime fear) when he created Wile E. Coyote. For the producers of *Duck and Cover,* however, the hot bomb is deployed enthymemetically (but for a brief flash, we never see or hear it) to create fear, and because the sound is invisible and inaudible, it is doubly frightening. The enthymeme is combined with the hot medium of documentary film that contrasts too sharply against the cool medium of the Bert the Turtle cartoon and the cold war. Hot and cold must be mixed carefully; otherwise, a film might produce reactions varying from hysteria to communist witch-hunting.

Fortunately, the transition from the Truman to the Dwight D. Eisenhower administration in early 1953 was marked by a downshift in the rhetoric of fear. Advised by experts in propaganda and psychology, President Eisenhower was made aware that Truman's efforts had succeeded in causing fear but that the fear caused was likely to lead to paralysis rather than action. Eisenhower was also wary of Senator Joseph McCarthy's witch-hunting. Eisenhower in both his public rhetoric and in iterations of government propaganda dialed down the fear so that Americans could be trained to react appropriately to an atomic attack.[24] That shift was also accompanied by thousands of air-raid drills staged from 1954 to 1961 that encouraged community building and, as hard as it may be to believe, fun.[25] Even so, *Duck and Cover* continued to be shown to American schoolchildren into the 1960s.

That the air-raid sirens encouraged community building is not a little surprising. Given recent scholarship about the rhetorical force of iconic images, this makes sense. In *No Caption Needed,* Robert Hariman and John Louis Lucaites develop a series of arguments about iconic images that can be applied to iconic sounds. The authors present six vectors of influence produced by particularly iconic images.[26] First, these images present "a pattern of motivation that can make some responses more likely than others." Second, images reproduce ideology and social order. Third, the iconic image draws upon and reproduces social knowledge that is learned simply by growing up in a particular culture. The fourth vector, the icon's ability to combine ideology with cultural assumptions, produces an even less conscious reaction. This for Hariman and Lucaites produces collective memory. Fifth, and critical to the visual scholars, icons model citizenship. Hariman and Lucaites's sixth and final vector of influence is the icon's ability to persuade. The air-raid siren evidences each of these vectors. The sound of the siren carries much meaning. It is instantly recognizable, and though it is polyvalent, its moan is always understood. It causes us to behave and think in directed ways. And it reminds us of our role as citizens in a larger society. In the 1950s, the alarms told us to survive so that we could live to fight another day. They told us to fear but not too much. They told us to obey and to support the American government against threatening others. Today, it signals us to remember a period of American history, to recall the cultural anxieties surrounding the possibility of doomsday, and to reminisce nostalgically about a "simpler" time when our parents and grandparents were strong and fearless. The air-raid siren is iconic and has been since the 1950s. It is so powerful in its sonic iconicity that it has transcended the senses and has gained a measure of iconicity as a visual image.

Illustrating a synesthetic overlap between images and sound, the air-raid siren is sometimes used in films visually rather than sonically.[27] What we remember from the *Blues Brothers (1980)*, for example, is not Elwood's voice announcing the band's upcoming performance but the ridiculousness of a retired police car with a massive air-raid siren emerging from its roof. Almost thirty years later, fans can still see the movie car as it visits auto shows and "Blues Brothers" performances across the nation. In *Forrest Gump* (1994), a panning shot at the beginning of the film contrasts a floating feather against a backdrop that includes an air-raid siren. The siren signifies much of what is to come in the award-winning movie. Among many significations, the siren foreshadows a shift from the audience's present (1994) and the protagonist's present (1981) to the 1950s, which is where the next scene takes us. The siren also warns of the many dangers that Gump will face and foreshadows the theme of war (the Vietnam War and the cold war) while reminding the audience of the importance of shelter, which is symbolized by the home Gump repeatedly returns to over the course of the film. It is not by accident that this powerful motif appears within seconds of the opening of the film. The air-raid siren is potent exactly because it is not used the way audiences have come to expect, sonically, indicating that particularly powerful icons transcend the senses.

During World War II, air-raid sirens produced identification between Americans and the British because these enabled radio audiences thousands of miles away to audibly experience the threat of German bombs. After the war, the air-raid siren produced a different kind of community; a community of American citizens ideologically committed to surviving so that their nation would emerge triumphant from a third world war. Though diminished, the sound of the air-raid siren has not yet lost its valence as an icon. The sound can still be heard during monthly tests and occasional emergencies, not to mention in popular culture in the form of films and music. Furthermore, the sound of the air-raid siren may retain its cultural valence even today because of related sounds that remind us always of the potential for bombs to kill Americans and destroy our way of life.

Bombs

While sounds related to the nuclear bomb are not as iconic as the sound of the air-raid siren, these do produce laughter, obedience, and fear. Wile E. Coyote, for example, is famous for his countless battles with gravity. As he attempts to capture Road Runner, Bugs Bunny, or the helpless wards under the protection of Sheepdog Sam, Wile E. manages over and over again to find himself high above a canyon with nothing between him and the ground but the occasional cloud. In his earliest appearance, *Fast and Furry-ous* (1949), Wile E. takes two long falls.[28] First, the coyote dons an "Acme Super Outfit" to catch the roadrunner. After jumping from a cliff, however, the coyote discovers that the outfit does not help him to fly. His first long fall is accompanied by the sound of a diving, propeller-driven airplane in a parody of a World War II dive-bomber and ends with a crushing noise. The second fall is precipitated by Wile E.'s effort to ski down a hill with a machine on his back that produces snow ahead of his skis. He overshoots Road Runner and finds himself making snow over a wide canyon. The coyote's second fall is accompanied by a combination of sounds, one of which would be separated out in later cartoons to great effect. This plunge features the sounds of descending violins, a dive-bombing airplane, and a falling bomb. But the sound did not work, and so Jones and his team of animators continued to tinker with it until after *Duck and Cover* began appearing in classrooms across America.

Beginning May 24, 1952, episodes of Road Runner and Wile E. Coyote feature the sound of a falling bomb without accompaniment suggesting that the cartoon's creators and, in particular, Jones and sound engineer Treg Brown, recognized the power of the sound in isolation.[29] By making one premise of the air-raid siren's argument audible (i.e., that a bomb was falling), Jones and Brown exposed the fear-mongering of productions like *Duck and Cover*. The constant use of the isolated sound of the falling bomb in the Road Runner cartoons of the 1950s was a parody of the fear induced by the government's manipulation of emotion through ubiquitous air-raid sirens and propaganda films.[30]

Jones understood the importance of scaring his audience during Wile E. Coyote's falls but not causing fear: "Ray Bradbury says that kids like to be scared, and you do. Psychopathic or psychological fright is one thing, but being scared is fun."[31] Unlike the federal government's psychological operations before Eisenhower became president, Jones wanted to reduce fear in favor of fun, the kind of fun moviegoers look for when paying to see a horror film, theme-park adventurers expect from roller-coaster rides, and audiences anticipate when listening to ghost stories. As Bradbury and Jones recognize, we all enjoy the spine tingling, the chill in our bodies, and the wide-eyed reaction to being scared. Wile E. Coyote's falls are scary yet funny. Under Jones's direction, the sound of the falling bomb does not signify death. Instead, it has become in popular culture the sound of someone humorously failing as when a guy "crashes and burns" after being rejected in his entreaty for a date or when someone is caught on tape performing a blooper (note the sound for crashing and burning should be of the rapidly descending plane, but it is of the bomb, which, of course, is intended to crash and burn).

As Jones wrote of Wile E., "In the Road Runner cartoons, we hoped to evoke sympathy for the Coyote. It is the basis of the series."[32] Jones describes Coyote as everyman. For example, sometimes everyman continues to pursue something not because of his original purpose but because of his stubbornness. Sometimes, he employs technology that lets him down. And it is the simple things that often undo him, like the thirty-five-cent element that failed in a half-billion-dollar unmanned space rocket.[33] The sound of Wile E. falling is a sound that we identify with. It is the sound of crashing and burning humorously but not of death.

Jones had aligned himself with Hollywood's left at least by the time he began organizing a strike against Walt Disney and his studios in 1941. By 1946, Jones was an active member of the Hollywood Independent Citizens Committee of the Arts, Sciences, and Professions, a liberal action committee that would soon be labeled a communist-front organization by witch hunters like McCarthy. In 1948, Jones joined Jerry Vorhees' congressional campaign as a poster illustrator. Vorhees' claim to historical fame is that he lost that election to a McCarthy ally, Richard M. Nixon.[34] Not surprising, cartoons with the Space Martian were explicit rebuttals of government policy during the cold war.[35] An analysis of Jones's work and, in particular, the use of sound in Road Runner and Wile E. Coyote cartoons reveals how Jones and his new employer, the film production and distribution corporation Warner Brothers Studios, protected the American psyche during the early part of the cold war and how norms of that culture were contested, even in the frames of a supposedly childish cartoon series.[36] Jones was aware of cold-war events and of government propaganda efforts, and he fought back sonically.

Once we read the noises closely, we can understand why government propagandists avoided using the sound of the falling bombs to which Jones turned in his cartoons. I pause here to read this sound carefully. We are familiar with the sound of a train and, in particular, of its rapidly descending pitch as it travels past us and its gradually decreasing pitch as it continues to travel away. Scientists refer to this phenomenon as the redshift character of the Doppler effect. In 1842, Christian Doppler determined that viewers visually experience a blueshift as objects rapidly approach. Because light waves are compressed when an object (like a star) travels toward the viewer, the object's appearance is shifted toward the higher visual frequency on the color spectrum, blue. And when objects rapidly move away from our eyes, we experience a redshift as the frequency of the light is stretched out, shifting the color of the object toward the less-compressed end of the spectrum, red. Three years after Doppler's discovery, Christoph Hendrik Diederik Buys Ballot tested the applicability of the theory to sound when he hired trumpeters to play music on an open railroad car as it passed a musically trained audience, who then described the change in pitch. Rapid movement away from a listener causes sound waves to stretch out, thus lowering their frequency, and, in terms of the musical scale, producing a sound that glides down in pitch. It is a sound that we have all heard—it is the loud car radio, the ambulance siren, and the ice-cream-truck tune that rapidly pass by.

How is this discussion relevant to the sound of falling bombs? We have been trained to believe that this sound is naturally occurring. It is a noise produced by laws of gravity, friction, atmospheric pressure, and so on. I do not here contest the naturalness of these laws. Rather, I contest the naturalness of the iconic sound of falling bombs. While these do make a sound determined by physics, and that is sometimes accurately projected in Hollywood films, it is a sound from the perspective of a particular listener: the listener away from whom the bomb travels. These are the sounds produced by a culture that has, since 1812, bombed others and not been bombed itself. Listen to a war film produced in Germany, and you are likely to hear a very different sound; the sound of something falling toward the listener has a gradually ascending or constant high-pitched scream, not an almost musical, falling whistle. The sound of the falling bomb that Jones made famous in the 1950s is the sound perceived by a people who are bombers and are not bombed. It is a sound of survival, not of death.

This, then, helps to explain why the sound of bombs falling away from Americans was never employed in government propaganda films. Those films succeeded because they prayed on the fears of a sound that was alien to all but Hawaiians and combat veterans, the sound of bombs falling toward Americans. Indeed, Americans still do not have that sound in their aural vocabulary. The sound of the falling bomb that Chuck Jones employed in the Wile E. Coyote cartoons could not be employed in films like *Duck and Cover* because it was the wrong sound, and it would not have induced fear. But it was employed to great effect in Jones's cartoons as a countermeasure to government fear-mongering. Jones hoped to scare his audience but not to cause them fear. It is not the sound of a dying coyote but rather the sound heard by the audience away from whom the coyote falls. Listeners survive the parodic attack against the bottom of the empty canyon because the listeners do not fall. Simultaneously, they can continue to identify with Wile. E. because he also survives, repeatedly, the sound of the falling-away-from bomb. The irony employed by Jones in his Wile E. Coyote cartoons may have contributed to Americans' critical skills and, in particular, their ability to combat fear-mongering. In turn, those critical skills may have made it possible to bring down McCarthy and his allies. It should not surprise us that the red-baiting senator had his career ruined in 1954 by a very sophisticated user of sound whose voice was nearly omnipotent, namely Edward R. Murrow on the show *See It Now*. Unfortunately, however, fear-mongering did not end with McCarthy's career. The history of the sounds of bombs was not yet over.

On September 7, 1964, a broadcast of NBC's *Monday Night at the Movies* revisited the sound of the Bomb when the program took an advertising break for its national audience. That break featured a new political spot from the Lyndon Baines Johnson for President campaign. Almost a half century later, the ad, known now as "Daisy," remains the most famous political advertisement in American history. The spot was produced by the firm Doyle Dane Bernbach (DDB), a firm founded in 1949 that grew up with television. DDB contracted out this spot to Tony Schwartz, who had established himself as a wizard of sound in a visual medium. In other words, Schwartz knew how to manipulate in ways beyond the ability of the vast majority of Americans to critically analyze. This ad marked the high point of his manipulations.

Daisy opens with a young girl pulling petals off a flower, the flower for which the advertisement is named. As she pulls the petals, the young girl attempts to count upwards to ten. Her fragile voice, her insecurity about counting correctly, and her failure to count properly all contribute to humanizing this young girl beyond any previous effort to produce sympathy for advertisement actors. Within fifteen seconds, the audience of *NBC's Movie Night* would have wanted to take care of this young child. Visually, the blonde girl is as cute as a button, and so the twinning of sound and image here works doubly to make her a sympathetic figure. When she ends her count at nine, the motion picture freezes. A serious yet metallic adult male voice begins a countdown over the frozen image of the girl as the camera zooms into the child's eye. At zero, we have reached the center of the eye, which transforms into documentary footage of a nuclear detonation and the rapid growth of a mushroom cloud accompanied by the sound of an explosion. Over the noise, Johnson poetically warns of the Third World War: "These are the stakes. To make a world in which all of God's children can live or to go into the dark. We must either love each other, or we must die."[37] Though the images are powerful, they only divert our attention from the real manipulation. The images paralyze; they do not persuade.

It is the sound of this commercial that carries tremendous rhetorical force. The sonic oppositions in the advertisement are many and intentional. A child's innocent voice, imprecisely counting daisy petals, is followed by the mechanical, militaristic voice of a missile countdown. The fumbling attempt to count from one to ten is countered with a precisely scientific countdown and then a stirring passage in the voice of the President of the United States of America. The sound of birds chirping gives way to the sound of an explosion. There is no irony or humor here. Nor do the noises and images rely on the sound of air-raid sirens and falling bombs; those are absent. Rather, this advertisement employs the sound of the Bomb exploding. The sonic and visual presentation of a massive nuclear detonation represents death, failure, and Armageddon.

In this advertisement, voices and noises hammer home the message that Barry Goldwater could not be trusted. As Chion argues, "Sound, much more than the image, can become an insidious means of affective and semantic manipulation."[38] We are compelled, for example, by the sound of the young girl's voice to empathize and by the twittering birds to think of nature. Her mangled words as words are meaningless. It is the voice that matters. That voice is a synecdoche for life, family, the future, and humanity. That the little girl also spoke with the same Boston accent as America's recently assassinated president would also have resonated deeply with listeners. Here was youth about to be cruelly murdered. The reverberating mechanical voice that counts down from ten to zero represents death, a countdown to the end and doom.

Again, numbers-as-words do not matter—it is the cold voice that takes us to the edge of our rational ability to think and to critique. Here is the voice of modernity and of killing machines run amuck. The voices provoke, according to McLuhan, sensuous involvement in the scene.[39] We are emotionally caught up in and logically disarmed by the sounds.

The greatest rhetorical force in this sixty-second spot, however, belongs to the most powerful man in the world. Lyndon Baines Johnson, President of the United States, almost universally popular after assuming the office upon the assassination of John F. Kennedy, spoke with the authority of both his homey voice and of the highest office in the world. Here was a man with his finger on the doomsday button. Here was God. This voice is *acousmatic*, a term borrowed from the Greek *akousma*, meaning "things heard."[40] For Chion, the acousmatic sound is the sound without a source that emerges from nowhere and occupies the space of everywhere; it is ubiquitous, panoptic, omniscient, and omnipotent.[41] These sounds have a haunting, ghostlike effect. Invisible broadcast media produce a distinct and potent kind of vocal presence. The *acousmêtre* is the most powerful of the acousmatic sounds because the term describes the sound of a voice without a body. Though Chion focuses on the phenomenon of the *acousmêtre* in the context of cinema, its effects also appear in television and even in radio where the invisible voices of the actors assembled by Orson Welles, for example, were able to cause panic across the United States in 1938.

We can now listen to the advertisement with an appreciation for its force multiplication. Not only is the voice that emerges from nowhere to occupy everywhere the voice of the most powerful man in the world but also the manner by which the voice is heard but not seen projects the acousmatic power of an all-knowing, omnipotent deity. As the Daisy ad's creator argues: "The electronic waves of the media suffuse the atmosphere we live in. McLuhan equated the media environment with the traditional definition of God, whose 'center is everywhere and whose margins are nowhere.' This relates to the phenomenon of sound. Sound envelops you, because it is spatial."[42] Schwartz's envelope is the presence described by Walter J. Ong and the *acousmêtre* described by Chion. Thus, not only is the sound that of the President of the United States and a sound whose center is everywhere and whose margins are nowhere, but the sound of the president's voice also wrapped listeners in a gentle cloak of sound, a poem, in an example of the sonorous envelope: "We must either love each other, or we must die." The voice in this advertisement simultaneously represented self, God, and father. What could be more powerful?

Not surprising, people responded viscerally to the sound of the advertisement. A *Chicago Tribune* reporter described the ad not by recounting its images but by focusing on its sound: "The one-minute spot shows a small girl picking daisy petals. She is counting, one, two, three, four, five . . . then a voice is counting down—five, four, three, two, one, zero—and an atomic bomb explodes. Next a recording of President Johnson's voice is heard. Then an announcer urges voters to support Johnson."[43] The sounds in the advertisement provoked a firestorm of protest. Republicans fought back hard against the message, claiming that the ad "was reckless and aimed at creating mass hysteria." One GOP representative worried that "children grow up with enough fears without seeing themselves blown up on TV while innocently watching Bugs Bunny or Walt Disney."[44] The ad had run during a nighttime showing of *David and Bathsheba*, a Biblical tale of sex and betrayal that few children would have been permitted to watch in 1964.[45] The complaint illustrates that Republicans understood the sonic connection between Wile E. Coyote and the little girl in the ad. Wile E. survived thanks to the ironic soundtrack of a triumphant bombing. In the Johnson commercial, Daisy died. The bomb had finally exploded.

Republicans also claimed that Daisy was a dirty trick because it employed an extremely negative caricature of Goldwater. John Burch, the chairman of the Republican National Committee (RNC), complained that the advertisement amounted to a libel and a "violent political lie" against Goldwater.[46] News outlets from the *New York Times* to the *Chicago Tribune* to the Associated Press to the *Washington Star* to *Time* magazine ran stories about the emotion-provoking spot, the first such political advertisement to get so much free publicity.[47] On October 2, Goldwater was still complaining about the spot even though it had run only once on the evening of September 7.[48] Because it was replayed on newscasts and repeatedly referred to during the campaign, the advertisement and its message reached a much-wider audience than the one watched by viewers of *David and Bathsheba*.[49]

What illustrates the ad's effectiveness is its enthymemetic argument. We have seen all of the words used in the spot: numbers running upward toward ten and back in a countdown along with Johnson's poetic passage and an announcer asking viewers to vote for the sitting president. There is no mention of Goldwater in the script, and yet virtually every voter who saw or heard about the ad understood that it was about the Republican candidate for president. Tony Schwartz wrote of it, "The commercial *evoked* a deep feeling in many people that Goldwater might actually use nuclear weapons. This mistrust was not in the *Daisy* spot. It was in the people who viewed the commercial. The stimuli of the film and sound evoked these feelings and allowed people to express what they inherently believed."[50] Elsewhere, Schwartz comments more generally: "People are most capable of receiving and understanding sounds they have heard before. A person responds most readily to sounds that evoke past experiences stored in his mind and available for recall. We hear a sound or word and associate it with similar sounds or words already experienced."[51] Though not a rhetorician, Schwartz understood the power of the enthymeme, an argument with a missing premise that the audience fills in by using their own preconceptions.[52] The Johnson campaign did not make the argument that Goldwater was a trigger-happy lunatic likely to blow up the world. Instead, the audience made that argument themselves. Attacks against the unfairness of the commercial were attacks against the very people whose votes Goldwater needed. Unsurprising, Goldwater lost the election by one of the largest margins in American presidential history.

Even without the use of an iconic sound, the air-raid siren, Schwartz was able to induce Americans into coming up with their own arguments, arguments that tens of millions of voters shared. That it was sound that induced the argument is difficult to refute. Schwartz, the wizard of sound, notes, "Commercials try to connect with built-in emotional signals stored in our brain. This is what enables the sound on radio and television to be 'subliminal' in the sense that those whom it influences are unaware of that influence."[53] The fear induced by the misuse of air-raid sirens in the 1950s created "built-in emotional signals" that were stored in millions of minds, waiting to be triggered by subliminal sounds. The nuclear explosion accompanied by the voiceover of God, in the form of Johnson's poetic crooning, provided that trigger. Plato, Kant, and their followers were correct in attributing emotional responses to sound. Music, voices, and noise emote, manipulate, and provoke in a manner that few people have ever been able to understand. But to dismiss sound as purely emotive is to ignore the power of sound to complete enthymemes and to persuade. Thinking critically about how sound manipulates enables us to approach sonic persuasion rationally and to understand its power.

The recognition that sound could be rationally employed to emotionally manipulate began in Nazi Germany and in the Soviet Union, where noise was first used to torture prisoners, an idea picked up by an influential American psychiatrist at the beginning of the cold war.[54] The United States government with the help of American academics further developed psychological operations that incorporated sonic torture and manipulation during the 1950s, resulting ultimately in its codification in a CIA interrogation manual titled

KUBARK (the cryptonym employed by the CIA during the Vietnam War), which was produced in 1963, the year before Schwartz employed sound to such powerful effect in the Daisy ad. Though interrogation techniques employing sound were supposed to be off limits after the Vietnam War, a recent study argues that these methods were still in use in 1976 at the School of the Americas, where the CIA trained right-wing military officers from Central and South America to carry forward American foreign policy.[55] Famously, sonic manipulation was employed to force Manuel Noriega from his home during America's 1989 invasion of Panama.[56] There, U.S. troops blasted American hard-rock anthems into the windows of the home to which the dictator had retreated. More-recent reports of the torture of Muslims sentenced to prisons at Abu Ghraib, Guantanamo, and other U.S. facilities indicate the ubiquitous use of sound to torture and manipulate prisoners.[57]

We will not pursue the use of sound as a torture device in American prisons, beyond noting that sound carries not only rhetorical force but physical force as well.[58] As anyone who has stood too close to an amplifier at a rock concert will tell you, loud hurts. More relevant is the military-industrial complex's turn toward sound as a device that could manipulate public opinion. The FCDA failed to use sound in a manner that would encourage proper responses, a failure addressed by Eisenhower after his election in 1952. And Johnson's presidential campaign employed a wizard of sound who turned to noises associated with war to help win the election of 1964. Militaries and politicians have been learning and, indeed, have gotten ahead of the curve in their ability to manipulate audiences who do not have the skills to respond critically. The lessons learned played out during the Persian Gulf War (1991).

Some of the most memorable images from that war are those of American air power. F-14 Tomcats launched into the air off aircraft carriers reminding Americans of movies like *Top Gun* and of the global reach of American power. Even more memorable are the video feeds from the nosecones of the bombs that pilots dropped over Iraq. Once launched, these missiles were remotely piloted toward their targets through video game–like controls. The grainy, black-and-white films illustrated to Americans not only their unbeatable military edge over the Iraqis but also their undeniable technological superiority. In the *Financial Times,* Peter Worthington states, "Thank goodness the U.S. has the military technology it has, because already it seems that in the air phase these video game missiles spare more lives than the mass bombings of another age."[59] Yet, studies taken after the Gulf War inform that these bombs did produce civilian casualties and made little difference to the military phase of the campaign.[60] Indeed, they were not even particularly effective at damaging Iraqi morale, indicating that the psy-ops function of the new high-tech and highly visible bombs was limited. Or was it?

The primary function of showing these videos was, like the *Duck and Cover* films, psy-ops. However, the psy-ops were exercised not against the Iraqis as a way of demonstrating overwhelming technical superiority but against the American people. The military-industrial complex employed the videos of bombings to prove to Americans two things. First, the videos illustrated in a tangible way the damage that American power and technology inflicted upon Iraq, even if the American people had no way of knowing if the damage was militarily significant. Second, and more important, the videos were employed to lead the American people into a false belief that their armed forces were not hurting civilians.

The American military-industrial complex had been working on guided munitions since World War II. During that war, inventors had discovered that aviators could navigate planes by using radio beams, the first indication that sound could be used to guide weapons. By Vietnam, the military had begun to rely on audio and visual means to guide missiles to their targets.[61] By the Persian Gulf War, those guidance systems, or at least those that the public were permitted to *see,* were visual systems. Yet, the GBU-15 missile was guided by both visual and sonic means. A pilot employed a camera in the missile's cone to determine whether the bomb was moving toward its target and if not, toggled a joystick to change the missile's trajectory.[62] But they signaled the missiles with radio commands. In other words, the missiles *heard* the commands of the military personnel who guided them toward targets. Yet, American audiences were not permitted to hear the missiles.

The message that the Pentagon stressed most repeatedly during the conflict was that they were not causing "collateral damage" (a euphemism for civilian casualties). Though not true, the videos enthymemetically presented evidence for that claim. Because Americans could not hear the sounds of the bombs they had been trained to fear, they completed the argument made by the videos by believing that there were no victims of those bombs. It "sounded" like the bombs were dropping on objects that were incapable of experiencing fear. This enabled the George H. W. Bush administration to maintain extraordinary levels of support, hitting an incredible 89 percent approval rating toward the end of a war that was waged to maintain the status quo, restore a repressive monarchy, and protect American oil contracts in Kuwait.[63] Airstrikes are often a political tool employed to manipulate public opinion and guarantee future contracts.[64] The camera-carrying bombs dropped over Iraq served an important function: convincing the American people that they were the good guys and preventing all possibility of empathy for the civilians on the ground. Though our government had trained Americans to fear the sound of bombs, they were deprived of their ability to hear the fear that Iraqis experienced.

The lack of sound made the destruction caused by American bombs victimless, thus the many analogies made in 1991 between the bombing campaign and video games. This metaphor suggests why a few journalists and academics recognized something was not right, even if they could not put their finger (or ear) on the problem.[65] Because we were not trained to read sound, we were unable to recognize the rhetorical force of the Pentagon's bomb videos. Cultural studies scholar Shehla Burney almost recognizes the sound of falling bombs as those dropped on colonized others: "The spectacular, slick and desensitized coverage of the war in the Western media as an exotic, exciting, star-wars video game with glib commentaries by immaculately groomed 'talking heads'—was an open text on the social and ideological construction of the same bloody war through the hegemony of North/South relations."[66] In 1991, imperial powers were once again dropping bombs on technologically deprived others. Like the Road Runner cartoons whose Wile E. Coyote is constantly set against the background sound of triumphantly falling bombs, the lack of sound turned the Pentagon videos into something other than reality. Burney continues, "What was totally missing in this high-tech media display was the blood and the tears of ordinary people—doctors, teachers, writers, intellectuals, carpenters—whose homes, schools, factories, lands and history were obliterated without giving peace a chance."[67] Hiding the images of blood and tears from the American public was accomplished by hiding the frightening sounds that would have prompted Americans to identify with and fear for the victims. Writing for the *Fort Lauderdale Sun-Sentinel,* Tom Kelly understands:

> If the kaleidoscope of violent images from the Persian Gulf had been subjected to review by some cosmic instant-replay system, the Great Zebra in the Sky might have reversed the original perception of many North Americans that modern warfare is a video game the good guys always win. Certainly that was the way it appeared back in 1991, when every night on television we could witness smart bombs entering the front doors of Baghdad buildings, cruise missiles following highways to their targets and Patriots blasting Scuds out of the skies over Israel and Saudi Arabia. One picture is said to be worth a thousand words. When the Pentagon edits the pictures, they can be worth billions, in both words and dollars.[68]

What Kelly almost gets is that the Pentagon has edited sound out of the pictures. Though the Pentagon might argue that the lack of sound was a collateral matter and that the video technology in the nosecones of its missiles never included sound equipment, one can and should ask why. Microphones are cheap to manufacture and produce; the missiles were equipped with devices that read radio waves. Was sound not included for financial reasons? During the Ronald Reagan years, the Pentagon enjoyed a budget that was nearly infinite, thus financial reasons did not prevent the inclusion of sound. Was sound not included for technological reasons? Recording sound is less complex than recording images. Was sound not recorded, then, for practical reasons? Here is where we get to the crux. The military might argue that sound was tactically unnecessary and thus not included. But, as we have discovered, the value of these weapons was largely psychological, thus a decision was probably made early on that sound would add noise to the clean-feed narrative of a successfully waged and nearly harmless war. Even if sound had been first included with the videos, the Pentagon would have found it necessary to remove that dimension before showing the films to the public. Sounds of falling bombs and explosions were not what Pentagon propagandists wanted Americans to hear. The sound of bombs dropping on human beings could not be tolerated. Instead, the videos presented to the American people had to be pleasurable and entertaining. They were employed to increase the popularity of President Bush, the military, and the policy of attacking Iraq.[69]

Kathleen Hall Jamieson has found in the writings of Aristotle an affect that she terms "empathematic."[70] This affect combines two powerful forces, the enthymeme through which audiences complete an argument with their own assumptions, and empathy by which audiences put their selves in the shoes of another. For Jamieson, Aristotle's discovery was that "audiences participate in creating the communication. You don't hand someone the message for maximum impact, you let the audience invest itself in the message." The soundless and grainy videos that looked like cartoon or video-game images, however, left American audiences of the Persian Gulf War with little possibility of stepping into the shoes of Iraqis and of completing the argument about the real effects of bombs. Those others had been turned into caricatures like Wile E. Coyote, always falling but never dying. Sound—and the Pentagon must have known this—would have risked turning Iraqi others into human beings with whom American audiences might empathize.

The Ticking Clock

The clock and its ticktock have also been tied to the Bomb. Incorporated into popular culture in *James Bond, Batman,* and *Get Smart* episodes of the 1960s, the sound of the ticking clock has come to represent a countdown to destruction, a bomb that is about to go off, and the superhero who is the only person capable of stopping the explosion.[71] In 2006, the RNC took advantage of the relationship between clocks and bombs when the committee produced a generic ad that was run on national television to support congressional candidates. The ad is so closely tied to the Daisy commercial that it is named after Johnson's first words in the 1964 spot: "These are the stakes." "The Stakes" borrows from both the theme of bombing and from the enthymemetic argument that Schwartz had made in his infamous commercial.[72] Like Goldwater, this ad extracted from the minds of viewers Vice President Dick Cheney's warning to Americans that Iraq and by implication Muslim terrorists would soon be producing mushroom clouds and further that Democrats would let it happen. No candidates are mentioned, nor are the parties (but for the legally mandated tagline "Paid for by the Republican National Committee"). Indeed, the 2006 ad betters its 1964 ancestor: it uses no spoken words at all until the mandatory tagline.

The 2006 commercial does, however, employ written words, 213 of them in one minute. Most of these are quotes from terrorist leaders superimposed on the images of Arab-looking men who speak without sound, the lack of their voices reminding us of the importance of denying their presence. Of course, the RNC hardly hoped that Americans would read all the messages dropped into the advertisement, so graphics specialists isolated phrases, faded the larger part of the texts, and expanded and brightened key passages: "kill the Americans," "inside America," "suitcase bombs," "nothing compared to what you will see next," and "what is yet to come will be even greater." This grammatical grab bag of phrases provokes an emotional response: fear.

Not surprising, the incoherent jumble of words mimics the phrasing of President George W. Bush, who critics have argued was terribly unpersuasive. Even though his sentences often make no sense, Bush's use of repetition and his emphasis on key words (like the brightening and enlarging of words done by the Stake's creators) hammers home his points. Indeed, the critics are wrong. Bush was incredibly persuasive. Nobody ever doubts his faith because Bush says "I believe" so much: "I know what I believe. I will continue to articulate what I believe and what I believe—I believe what I believe is right." It is not just the repetition, however, that makes Bush persuasive; it's the mangling of the language, a mangling that is so bad we can only draw one thing out of the passage above: "I believe." And because the spoken words pass so quickly, listeners are denied the possibility of thinking critically about the mess. Thus, as Jay Heinrichs informs us, when Bush says, "Part of the facts is understanding we have a problem, and part of the facts is what you're going to do about it," audiences only get "facts . . . understanding . . . problem . . . facts . . . do," and the President comes across as scientific in his analysis of the problem.[73] Similarly, if one reads the highlighted terrorist quotes in the Stakes, one is likely to come away with "kill . . . America . . . bombs . . . next . . . even greater." That's frightening!

A series of ironies helped to compose the spot: in 2001, Bush asked network and cable news broadcasters not to use video messages from Osama bin Laden and his allies. Of course, the RNC resorted to those images and messages in 2006. Because they were fresh (news broadcasters had largely refrained from providing terrorists with a forum on American television and radio), the words were that much more powerful and, to extend the irony, that much more worthy of being rebroadcast as news.[74] Indeed, Al Qaida could not have done a better job of marketing themselves. Furthermore, those who might have thought critically about the ad may have recalled that after five years under a Republican president and four years under a Republican Congress that claimed they would get the world's most wicked terrorist, bin Laden was still making videos.[75] But few Americans were able to think critically about the advertisement because the frightening words were not the key to the spot's persuasive power.

Why did audiences, political commentators, and television executives recoil from the ad? Because of its extraordinary sonic manipulations. For the first forty-five seconds of the commercial, the only sound we hear is that of a ticking clock that makes two sounds (ticktock) per "second." But it's no ordinary clock. For one thing, the time between ticktocks is only .58 seconds long (though the intervals remain consistent), speeding up the pace of the video, not to mention listeners' heartbeats. With each tick, the sound grows louder from the open to the thirty-second mark, beginning at a barely audible amplitude and ending with fifteen seconds of ticking that is loud enough to be annoying. The purpose, of course, is to create an ominous atmosphere of looming catastrophe; something is about to happen. The clock is counting down to Armageddon. The sound of a ticking clock focuses our attention on time, and because that sound has been used so often in popular culture to signify a threat to life, when the RNC paired it with images of terrorists, they knew that Americans would create from their assumptions an argument about time running out and Armageddon: "kill . . . America . . . bombs . . . next . . . even greater."

So far, we have an effective sonic device for manipulating emotions and, in particular, for stoking fear and a sense of impending doom. But this isn't what makes the ad special. It's the forty-five-second mark of the ad that works extraordinary rhetorical force. At second forty-five, the ticktocking stops. There is silence, and for the first time, a real second occurs (actually 1.4 seconds). Like FDR's implied question, "What do we have to fear?" that pause drags out the anticipated destruction. Sonically, the shift from loud ticktock to silence has also changed the beat of the commercial. The next thing that happens is a simultaneous shift in image and sound. On the screen, we see an explosion, and we think we hear an explosion; only the sound that we hear isn't an explosion at all. It is the sound of a beating heart that repeats until words appear: "These are the stakes. Vote November 7th." Though a Democrat, Schwartz would have been proud.

Incredibly, the advertisement employs the memory of the Daisy advertisement by quoting directly from the 1964 commercial ("These are the stakes"), by redeploying the verbal countdown as a sonic ticking, and by alluding to Cheney's warning about mushroom clouds. Thus, the Republicans associate their candidates and their message with that of Johnson, who successfully won his election and defended America against a similar threat from communist bombs. Though Schwartz may have thought it difficult to top the sound and image of a cute girl awkwardly counting daisy petals, the Republican National Committee bests the Democratic commercial through the acousmatic projection of the sound most closely related to the sonorous envelope: the heartbeat. We do not see it, of course, but the sound we hear is the sound of life in general and of our own life in particular. Furthermore, by employing the memory of the Daisy advertisement and our social knowledge of the relationship between ticking clocks, heartbeats, and nuclear bombs, the RNC enables the audience to make their own argument that America will soon be attacked by terrorists with the doomsday device.[76]

Not surprising, Democrats assailed the ad, providing the commercial with an even-wider audience, who, because of the controversy, got to see it on national news outlets like CNN, where it was broadcast again as news.[77] However, the advertisement was probably broadcast too late (October 22) to gain as much attention as Republicans would have liked. Additionally, after a knee-jerk reaction, Democrats realized that to continuously denounce the ad as Goldwater had done with the Daisy commercial would only have revivified the commercial into an endless replay loop on the cable networks. So Democrats quickly dropped their protests and shifted their rhetoric instead to reminding voters that bin Laden was still alive as evidenced by the Stakes, the only effective rebuttal they could make.

Without the ability to critique sound, we leave ourselves open to the awesome manipulative powers of Madison Avenue professionals, the military-industrial complex, and political operatives. They have learned to master sound in a way that overpowers our ability to think rationally. It is why corporations like NBC employs three chimes, pitched in the scientific designation as [G_3], [E_4], and [C_4] (in the key of C) to distinguish its network from the others. It is why the Pentagon does not include sound with its videos of bombs dropping on Iraq. And it is why advertisements like Daisy and the Stakes compel voters to act. Lawrence Levine argues that we can read sound. I argue that we must read sound.

6

On Sound Criticism

Thomas A. Edison invented the phonograph in December 1877 and promptly put further development of the invention aside as he perfected another innovation, electricity. Edison only made the phonograph practical in 1888. Though Edison made few recording devices that first decade, stories abound about the things that Edison recorded. The most famous of these tales is that Edison sang "Mary had a little lamb" into his first machine in 1877, though this is likely apocryphal. Primary texts indicate that Edison or, likely, one of his assistants shouted "Mad dog" into the horn and then played the message backwards, projecting a reversed voice out of the device: "God damn!"[1] That recording, if real, has long been lost.

Only two recordings that were supposedly made in 1878 are still playable. The first, and most certainly from 1878, is a recording of three passing locomotives. Edison and his assistant Charles Batchelor were tasked with responding to complaints about the noise made by the New York City Elevated Railroad Company. It is remarkable that the study of the noise pollution caused by machine-age technologies should have been one of the earliest Edison recordings, illustrating that those who witnessed the rapid introduction of elevated trains, steam engines, and factories were greatly concerned about the intrusion of modern sounds into their lives.[2] Today, we tolerate these noises and have largely learned to block them out of our consciousness. In 1878, those noises could not be ignored and produced a drive to study how to reduce them. Thus, the inventors recorded a phonautogram (a graphic depiction of sound that is visually read) of the noise of three passing trains. Recently, sound engineers have digitally turned that image into readable noise.[3]

The other purported recording from 1878 is that of a man, perhaps François "Frank" Lambert, speaking the hours of the day as the cylinder moves both forward and backward. This recording has lasted the years because it was cut into a cylinder made from lead rather than tinfoil. And yet, it may not be quite as old as its champions claim. Having dated the recording to 1896, a group of scholars involved in the First Sounds project (the mission of which is to recover and study early sound recordings) has challenged the collectors of the Tinfoil project (enthusiasts who restore and digitize antique recordings), who claim the older date. If the recording is from 1896, that makes it old but not particularly interesting for the First Sounds scholars or the enthusiasts because time has left intact plenty of playable recordings from the 1890s. But the controversy over the recording's age makes Lambert's cylinder doubly interesting for a study about reading sound.[4]

Lambert had immigrated to America from France in 1876 ("Frank" is an anglicized version of his given name) and found a job with a firm that was contracted by Edison to design a talking clock based on the recent phonograph invention.[5] Edison appears to have wanted to replace the disturbing sound of alarm bells and hourly chimes with a voice because that sound is less disturbing than dinging bells. It appears that Lambert successfully invented the proposed talking clock in 1878 or at least was partially successful; he succeeded in producing a recording and playback machine. Whether it worked is part of the controversy. In 1896, Lambert was asked to demonstrate his 1878 machine as evidence for one of the multitudinous lawsuits about patent rights that stymied the recording industry until 1919. But Lambert claimed that he had lost the device of 1878, causing the court to demand that he make a copy of the original machine, which in turn prompted Lambert to ask for compensation. Scholars argue that the extant device is the copy of 1896 (though Lambert claimed to have destroyed it soon after he had shown it to the court). Enthusiasts claim that the original device was never lost. Perhaps, Lambert, in his sworn deposition, was trying to protect himself from a lawsuit, as he had not passed his invention along to Edison. Indeed, Lambert was at least a little duplicitous, having attempted to extort a reward for not damaging Edison's patents. Though evidence from patents and Lambert's sworn testimony indicate that he lost the original machine, the documents are baffling because they cast as much doubt on the scholarly theory as they do on the enthusiasts' date. What we know is that a machine exists, and that regardless of Lambert's claims, it is very likely either the original machine of 1878 (which he claimed to have lost) or the copy of 1896 (which he claimed to have destroyed). Because it is possible to read sound closely, perhaps the recording itself can tell about its origin.

At 7.5 seconds into the recording, a voice begins to emerge from the fuzz; however, the pitch and speed of the voice in addition to the buzzing make it virtually impossible to understand. After adjustments to speed, pitch, and noise are made using sound-editing software, a series of spoken hours beginning with "eight o'clock" can be heard. No adjustments are needed once Lambert reaches "four o'clock," at which point the recording revolves at a consistent eighty-five revolutions per minute (rpm). Elsewhere, the recording winds faster and must be slowed down or runs slower than eighty-five rpm and must be sped up. Lambert handspun the cylinder by a crank, thus making it possible to record and play back at varying rates of speed. Unfortunately, in order to get the most precise playback, the recording must revolve at the same rpm that Lambert employed over a century ago. This is a near-impossible task because he cranked that machine at a wide variety of speeds.

At the beginning of both ends of the recording, rpm quickly increases as Lambert starts up the machine. That both ends of the recording have a rapidly increasing revolution rate informs us that the recording was made in both directions. Thus, the mumbling at the end of the forward recording is a backward recording of a voice announcing hours. On the way back along the cylinder, starting at three seconds in from the end (or beginning) of the recording, the listener can make out Lambert's voice again running through the hours from "three o'clock" to "six o'clock," given adjustments to speed, pitch, and noise. Like the forward track, the hours shouted into the machine on the back track also reemerge at varying rates of speed.

The sounds tell a few things. The significantly varying rates of speed must be intentional. Though it is likely that Lambert could not have kept the crank revolving at a perfect rate of eighty-five rpm over thirty seconds (after all, this was an early effort), he could at least have maintained a consistent rate of speed that would have produced an audibly legible voice. Instead, judging by the sound, the rates of speed vary greatly during the seven to twenty-two seconds of the forward track; the rate then settles into a consistent speed near eighty-five rpm from the twenty-two- to thirty-five-second marks. If this was a prototype model, given the newness of the invention, Lambert may have been testing to see what his device could do. It is also likely that the inventor would have had to vary the speed of the machine to determine the appropriate rate for recording and playback. More practical, for this talking-clock machine, in particular, if the entire cylinder was going to be used to measure time, then Lambert would have had to know where along the cylinder he needed to record his voice and at what speed he needed to turn the crank. Lambert needed to sync the cylinder with time. Lastly, and doubling the complexity of the task, the inventor appears to have determined that rather than require talking-clock owners to reset the recording after every twenty-four hours, the recording would automatically reset itself by playing the first twenty-four hours forward on the cylinder and the second twenty-four hours backward in a never-ending loop or, at worst, a clock that only had to be reset once each day—an ingenious solution.

If Lambert had created a second prototype in 1896, there would not have been a need for him (other than authenticity, which he did not need because he claimed to have destroyed the original) to record at variable rates of speed and both backwards and forwards. He only needed to show that he had a viable device that could record and play back. Indeed, if his claim to have invented the first nontinfoil recording device were to have taken precedence, then it would have behooved him to make the sound clean and legible, thus proving that he had created a working machine rather than a device that produced gibberish. Furthermore, a rough recording would also have subjected Lambert to questions about his claim to have reinvented the device at the court's order, as Lambert hoped to claim damages for the efforts he supposedly went through to re-create his invention. Given that the purpose of the purported re-creation was not a talking clock but evidence in a patent case about the origin dates of previous inventions, the voice on the 1896 recording should have emanated from the machine more legibly. Lastly, when Lambert ends with the forward sequence, he finishes enthusiastically and triumphantly: "Twelve o'clock↑" with an up note like the end-of-sentence declamations common to the orotund style. By 1896, that style was beginning to fall out of favor. These arguments lend credibility to the enthusiasts who claim Lambert's recording was made in 1878.

Conversely, given that in 1878, speakers were still employing the orotund style in which *r*'s were usually rolled, it would be unexpected if a newly transplanted Frenchman or even an English-speaking assistant did not trill his *r*'s. The *r* in the phrase "four o'clock," which is audible on both the forward and backward tracks, does not display the trilling of French speakers, Shakespearian actors, and orotund presidents. Though difficult to hear, the *r* in "three o'clock" on the backward track is also not trilled. Any speaker in 1878 recording a series of sounds for an expensive device designed for consumption by elites would have been expected to roll *r*'s. By 1896, Lambert or his assistant may have avoided the trilling to conform to the gradual shift that eliminated the orotund style from the recordings that were played in the saloons of working-class Americans and parlors of middle-class Americans. Thus, the untrilled *r*'s suggest a date of 1896 for the machine.

In terms of settling the dating dispute between the enthusiasts and scholars, the two pieces of sonic evidence—the variable rates of speed and back masking, the declamation, and the lack of rolled *r*'s—contradict each other and leave additional questions about the value to historical inquiry of reading sound. The sound provides evidence for each of the competing claims. Yet, the sonic evidence mirrors the verbal evidence, which is similarly contradictory. The contradictions seem very much related to the manipulations of Frank Lambert. He is a character who deliberately muddied the waters and made himself difficult to recover in both the verbal and sonic records. Though reading sounds cannot help us to date the device any better than the written record can, it does at least inform about Lambert's efforts to profit from his inventions and duplicities.

What is most interesting about this recording is not the inventor known as Frank Lambert or even the purported date of the invention, given that sound engineers from the First Sounds project recently produced legible recordings from 1860 that make the 1878/1896 recording largely trivial. What makes the recording notable is Lambert's manipulation of time. In the first place, this talking clock should precisely inform listeners of the time. Yet, the recording makes it difficult to time its invention with any kind of certainty. Both 1878 and 1896 are plausible dates for the cylinder. Additionally, Lambert made the recording using a variety of rates of speed, alternatively slowing down and quickening the speed at which he turned the crank and thus the time at which each hour is spoken. That a voice clock would crank at a variety of speeds is mind-boggling, for the very usefulness of the clock relies on the mechanism turning at extremely consistent rates. Clocks are supposed to sound: "ticktock ticktock," not "tick . . . toooooooock tiiiiiicktock!" Lambert has warped time, multiplying the deceptions. Not surprising, in his pronouncements from "eight o'clock" in the morning through "twelve o'clock" midnight on the forward part of the recording, Lambert skips from the morning "twelve o'clock" to "four o'clock" with one garbled word between them. The intermediary word marks the moment when Lambert shifts from cranking the machine at variable rates of speed to a consistent eighty-five rpm. The multiplication of confusion continues when Lambert's evening sequence skips "ten o'clock."

Sound recordings can produce the impression of presence out of time. Like the journalists who responded to press releases about Edison's new invention by writing about how the phonograph would keep the voices of the dead and like the listeners who compared hearing phonograph recordings of the dead to séances, recordings disturb chronologies.[6] Though Robert Browning and Theodore Roosevelt are dead, they still speak to us. If the voice is presence, then these historical figures are present in a manner that only sound can convey. How many times have we heard Frank Sinatra or some other famous singer called timeless? But recordings of their voices are not timeless; if we listen closely, we can discover motifs particular to a period. Sinatra attempted to sound like Billie Holiday in the early 1940s and sang accompanied by electric guitars after these became popular in the 1950s. And yet, he still speaks to audiences of romance, heartbreak, and joy. Recorded sounds like Sinatra's voice may not be timeless, but they do warp time.

The warping of time is particularly evident in the Doris Day tune titled "Tic, Tic, Tic," which accompanied the 1949 film *My Dream Is Yours* (which combined live action with Bugs Bunny). In a light and happy style, Day sings lyrics that conflate nuclear radiation with clock time and with the quickening beat of her heart.[7] It is a love song.

> You're such an attractive pick
> You give me a radioactive kick
> It's distractive the way you stick
> But love, love makes me tic.
> I tic, tic, tic and my heart beats quick.
> How can anything go wrong?
> When I'm listening to that Geiger counter song
> I tic, tic all day long.

Set against the background of tonally ascending ticking on a variety of instruments, Day associates the sounds of radioactivity, the atomic bomb, the clock, the heartbeat, and love. Given its relation to radioactivity and Armageddon, the upbeat vibe of the song is discombobulating. However, at the time of the film's release (April 16, 1949), the Soviet Union had not yet acquired the atomic bomb, which when tested August 29, 1949, caught U.S. intelligence agencies and Americans by surprise. Before then, the atomic bomb was a triumphant innovation that would keep Americans safe, and Geiger counters measured the fallout of America's nuclear program, rather than fallout from enemy attacks. Today, of course, the conflation of nuclear radiation and love is impossible to reconcile. Even listening through an historian's ears, the song is hard to believe. It simultaneously warps time and is out of time.

The ex-slave recordings of Phoebe Boyd are likewise out of time as illustrated by a flaw in the recording. Soon after the interview begins, the disk stops abruptly. There is a gap in the historical record between the moment the disk has stopped recording and the moment it begins again. When it is restarted, we hear all five speakers who were present for the interview. The scholars and interpreters attempt to reassure Boyd, who seems to have fallen into the recording horn because her hand slipped off the side of the device. Though Boyd is an octogenarian, the voices tell her that everything is "all right" as if she is a child. The episode reminds us that the only reason she is being recorded is that she speaks from the era before electricity and the phonograph. As much as her dementia causes her to lose track of time and her old age causes her to shift from speaking from seventy years in the past to the previous day, it is the technology and the purpose of the recording that are chronologically out of sync. Incredibly, she is present for Abraham Lincoln's presidency, Franklin Delano Roosevelt's presidency, and Barack Obama's presidency. Here, we have Lambert's clock and Doris Day's tune all over again.

Listening closely to the sound of these recordings—through close reading and through theories like psychoanalysis—permits us to witness history in new ways. These recordings make the past present in a manner that disturbs our chronologies and yet illuminates our understanding of history. By applying to sound the techniques of reading that have been used to draw meaning from words and images, we open up a new channel to our ancestors: the channeling of the dead via the ear canal. All sound recordings, after all, are live.

Methods

Can we read sound? Of course. Yet this may have been easier to do during the 1930s when theorists like Rudolph Arnheim were grappling with the effects of radio, though even then, sound was difficult to read. His contemporary Theodor Adorno believed that radio and popular music encouraged listeners to stop thinking.[8] Perhaps we are compelled to hear rather than listen to music because musicians over the course of the twentieth century became increasingly precise in their timing—a machinic precision that turns music into the sound of mass production and consumption.[9] Employing the difference that Kevin DeLuca poses between the gaze and the glance, rather than listen critically to sound and in so doing employ sound agentically, modern media, according to Adorno, has left us with the ability only to hear. We are trained not to consciously question consumerism, hegemonic capitalism, or the culture that embeds us in its sounds. Instead of using sounds (listening), we are used by sounds (hearing).

Though obviously a form of intentional human communication, even music has proven difficult to read.[10] Gilles Deleuze and Felix Guattari argue that with musical precision—the dogma of meter—comes an inability to think.[11] Because music has increasingly become a precise, industrial, mass-consumed product, the philosopher Allan Bloom complains of its effects by recourse to Plato's ancient diatribe and Kant's more recent polemic against music: "Music is the soul's primitive and primary speech and it is *alogon*, without articulate speech or reason. It is not only not reasonable, it is hostile to reason."[12] Philosophy has long posed an opposition between words that convey reason and music that produces unreason (*alogon*). That does not stop music and sound from being persuasive, however. Rather, *alogon* should draw the attention of critics in order to disturb its machinations. If we recognize the dogma of meter, then we can rationally critique the effects and affects of that dogma.

In 1932, poet, playwright, and critic Bertolt Brecht argued for a counter to the reproduction of sounds: "The increasing concentration of mechanical means and the increasingly specialized training . . . call for a kind of resistance by the listener, and for his mobilization and redrafting as a producer. . . . This exercise is an aid to discipline, which is the basis of freedom."[13] Like the workers who yelled over the factory floor in the 1930s to drown out the machine noises and make the time go by, Brecht calls upon us to talk back to the sounds produced by radios and phonographs, as well as iPods and computers. Only by understanding and by becoming a producer of sound can we free ourselves from the forces of mechanical reproduction. Deleuze and Guattari likewise argue that it is in questioning rhythm where we can find the ability to think critically. Not surprising, in his *Manifesto of Futurist Musicians*, Balilla Pratella opens the twentieth century by demanding that composers substitute free verse for metric structure.[14] If we analyze the precision and mechanization of sound, we can discover how it reflects the capitalist superstructure. Though mass-reproduced music, as Adorno warns, is not designed to solicit a critical reaction, it does not prevent us from acting. We can read sound.

How then can one read sound or, to use Lawrence Levine's word, listen? On the one hand, it is as simple as reading the score of a composition with its quarter notes, sharps, and clefs. Indeed, music, as much as mathematics, is a universal language. Yet, musical notation may actually prevent scholars from reading sound critically.[15] Listening is much more than learning to visually read a new language. Rather, it involves learning a new way to read, one that interprets and produces meaning through the ears rather than through the eyes. The task is far from impossible and indeed has been done, in bits and pieces, before. Because we have learned new ways to read religious texts through exegesis, because we have learned new ways of reading literature through what critics term close textual analysis, because we have learned new ways of reading entire cultures through what anthropologists call deep reading and thick description, and because we have learned new ways to read visual objects, a sustained effort at discovering how to read sound should produce remarkable discoveries that illuminate histories and cultures.

In learning how to read sound, one can start from common sense. As with images, we are pretrained to read some sounds because of our genetic wiring and our experience of the physical world. This inheritance enables us to read the sound of a yell as that of someone who needs help or is hurt or angry. Similarly, we may read the sound of coughing as disease and, thus, as potentially dangerous. Likewise, we do not need to see heavy rain to know that we should not go outside. A tree that creaks in a long glissando threatens to fall and warns us to get out of the way. Significantly, this warning does not come from the tree, as the tree has no consciousness and cannot exercise agency. Rather, it is the listener, who interprets the sound and, thus, creates the argument. That is often how sound works persuasively. Though a yell or a cough may convey "danger," though rain may convince an individual to stay inside, and though trees may warn to get out of the way, it is the audience who produces the argument.

That being said, sound can be read for more signs than those produced by listeners. By the early twentieth century, psychologists had learned to look for deeper sonic clues as well. A quiet voice, for example, might mean shyness or dishonesty. Often, the speaker might not even know that he or she is conveying such information. Our cultures also provide us with interpretive guides to sound through *habitus,* which is habituation to the connection between a specific sound and a specific act. If I hear an ambulance as I am driving, I pull to the side. I have been trained to understand that the siren means make way for an emergency vehicle. Similarly, the wail of an air-raid siren means find shelter or that it is the first Tuesday of the month. This kind of meaning is not a recent invention. R. Murray Schafer, Alain Corbin, and Mark Smith, for example, recognize that church bells have for centuries created the sonic environment of a community.[16] Those bells convey a variety of meanings: time, place, God. Furthermore, though they may not know it, individuals connect the space of nations and/or communities through noises, like a language or dialect, the lowing of cows, the whine of lawn mowers, or the Islamic call to prayer, that are particular to the nation or community.[17]

Because we do not read sound, we are often manipulated by culture producers. By the turn of the twentieth century, for example, Impressionist composers had learned to employ musical instruments to mimic the sound of the everyday as exemplified by Modest Mussorgsky's *Pictures at an Exhibition* (1874), a musical piece that brings paintings, sculptures, and whole galleries to our aural attention. Critics even argue that music can represent "a storm more effectively than the actual recording of thunder and lightning, conveying the emotional effects that often accompany such an event."[18] Teachers of rhetoric have for millennia encouraged speakers to manipulate their voice to portray passion, calmness, and authority. Listen to the voice-overs on biographical and negative political ads, and you will hear twenty-five-hundred years worth of pedagogy. Corporations are also aware of the power of tone in the voice and of the usefulness of ditties and musical themes. Government, too, has discovered the powers of noise and the voice. The Nazis and Soviets, for example, began to experiment with sonic torture in the 1940s. And the next time you hear an automated message at an airport or on a subway train, consider asking yourself why the voice is male or female. The answer is not "random."[19] Here, again, however, it is often the audience who interprets the sounds as persuasive arguments. The audience interprets the symphonic production as a tempest. Auditors believe that the speakers know what they are talking about because of the tone of their voices. Listeners send chills down their spines as they make sense of the screeching violins in Alfred Hitchcock's film *Psycho*.

In the 1920s, the composer Darius Milhaud attempted to combine advances in impressionist music with current psychology when he wrote about a black blues singer "whose grating voice seemed to come from the depths of the centuries sung in front of the various tables."[20] Milhaud was on to something. Thirty-five years later, James Baldwin penned "Sonny's Blues," a short story that mined the depths of black frustrations and anxieties in a nation long dominated by Euro-Americans. Baldwin wrote of this genre of music that had been played by black artists since the Civil War:

> Then they all gathered around Sonny and Sonny played. Every now and again one of them seemed to say, Amen. Sonny's fingers filled the air with life, his life. But that life contained so many others. And Sonny went all the way back, he really began with the spare, flat statement of the opening phrase of the song. Then he began to make it his. It was very beautiful because it wasn't hurried and it was no longer a lament. I seemed to hear with what burning

he had made it his, with what burning we had yet to make it ours, how we could cease lamenting. Freedom lurked around us and I understood, at last, that he could help us to be free if we would listen, that he would never be free until we did. Yet, there was no battle in his face now. I heard what he had gone through, and would continue to go through until he came to rest in earth. He had made it his: that long line, of which we knew of Mama and Daddy. And he was giving it back, as everything must be given back, so that, passing through death, it can live forever. I saw my mother's face again, and felt, for the first time, how the stones of the road she had walked on must have bruised her feet. I saw the moonlit road she had walked where my father's brother had died. And it brought something else back to me, and carried me past it, I saw my little girl again and felt Isabel's tears again, and I felt my own tears begin to rise. And I was yet aware that this was only a moment, that the world waited outside, as hungry as a tiger, and that trouble stretched above us, longer than the sky.[21]

Not only does music under the pen of Baldwin have the potential to mimic thunder but it also has the power to bring back history, to stir memory, and to return the dead to existence. By reading sound closely, Baldwin heard the dead and revivified a long passed culture. In music, there is much meaning.

The introduction of sound-media technologies that have been adopted by billions of people around the world—devices from the phonograph to the iPod, radio, the "talkies," television, and computers—has provided some individuals with a newfound ability to persuade. One sound technician notes, for example, that sound effects are "the art of painting pictures for the imagination."[22] Already in radio shows of the 1930s, the organ was employed to punctuate stabs, imitate gunshots, announce screams, play doom chords, intone somber marches, and illustrate staccato frenzies.[23] During the opening of the radio program *The Weird Circle* (1943–45), the announcer asks, "Bell keeper, toll the bell so that all may know that we are gathered again at the weird circle."[24] In doing so, the show's producers sonically create a community of listeners, borrowing meaning from ancient church bells while employing a very modern technology to reproduce the effect of the campfire circle and the impression made by a good story.[25] The laugh track that studio executives introduced to television in 1950 is yet another device intended to create meaning, in this case that the line recently uttered by an actor was intended to be funny. These examples provide insight into how media corporations manipulate sound to persuade audiences to listen, to participate, and, most important, to buy the sponsors' products.

Scholars have grappled now and again with reading sound that we normally hear rather than listen to. This impulse has a long history. Reflecting how industrial noises become invisible as cultures adapt to them, Henry David Thoreau performed a lengthy reading of the sound of a passing train that would probably be impossible to replicate today. In the twenty-first century, we might hear a train whistle and think, "That's a train whistle," hardly a thoughtful analysis of a sound. When steam engines on steel rails were new, however, Thoreau wrote a critique of the iron horse he heard chugging through the forest near Walden Pond. Writing about trains in the years immediately after their introduction to the Massachusetts woods, Thoreau did not hear them as white noise. Instead, he was forced to listen.

> The whistle of the locomotive penetrates my woods summer and winter, sounding like the scream of a hawk sailing over some farmer's yard, informing me that many restless city merchants are arriving within the circle of the town, or adventurous country traders from the other side. As they come under one horizon, they shout their warning to get off the track to the other, heard sometimes through the circles of two towns. Here come your groceries, country; your rations, countrymen! Nor is there any man so independent on his farm that he can say them nay. And here's your pay for them! screams the countryman's whistle; timber like long battering-rams going twenty miles an hour against the city's walls, and chairs enough to seat all the weary and heavy-laden that dwell within them. With such huge and lumbering civility the country hands a chair to the city. All the Indian huckleberry hills are stripped, all the cranberry meadows are raked into the city. Up comes the cotton, down goes the woven cloth; up comes the silk, down goes the woollen; up come the books, but down goes the wit that writes them.[26]

Would that we might listen to the train in the same manner as Thoreau. But our ears have become so accustomed to cacophony and the industrial sounds of modern life that the noise of the train has become almost meaningless, a Musak-like background to our daily existence. Before noise "pollution" became the soundtrack of Western life, the Italian Futurists celebrated this cacophony, illustrating again the difference between listening and hearing. Reveling in the new sounds of industry, automobiles, and machines after the turn of the twentieth century, Luigi Russolo trumpets the sounds of modernity: "Ancient life was all silence. In the 19th Century, with the invention of machines, Noise was born. Today, Noise is triumphant and reigns sovereign over the sensibility of men."[27] Indeed, it has so triumphed that it now invisibly operates through broadcast media and background noise to manipulate our emotions and shape our identities. If we are to understand these manipulations, we must force ourselves to listen again.

The best work on sound has consistently been written from within the Marxist tradition as exemplified by Adorno and Jacques Attali. Karl Marx provided a way of thinking about music as a product and how that product's manufacture, sale, and consumption produce or reflect alienation, the division of labor, false consciousness, and so on. Adorno was of all scholars most deeply informed by and curious about sound and, not surprising, his primary interest in sound was materialistic (though he extended his research into phenomenology). Adorno asked and answered how jazz, phonographs, and religious radio broadcasts affected culture, particularly through production, ownership, and atomization. Occasionally, Adorno ventured more deeply into sound itself, in particular when he described the impact of Beethoven. Related to Marx's concept of the alienation of labor, Adorno's read of Beethoven's late works enables him to describe how the Seventh Symphony, by reconciling and synthesizing the subject and object, reflects the impact of the Enlightenment on the division of the subject (the composer) from the object (the symphony) and, thus, the production of selfhood.[28] Attali's key work on sound, *Noise: A Political Economy of Music*, hews much more closely to the original Marxist line. Attali's thesis is that noise announces the future in a sonic, Hegelian dialectic. As he discovered, "what is noise to the old order is harmony to the new."[29] Along these lines, post-Marxism has produced one of the most sophisticated readers of sound yet. Jonathan Sterne has researched deeply into the practices of listening (particularly in his book *The Audible Past*) and the power of sound to manipulate mood, especially as it relates to the purchasing of consumer goods that one might think are unrelated to sound.[30]

The Marxists are best at reading music. Adorno and Attali study Beethoven and popular music, respectively. Also employing Marx, Lewis Mumford, in *Technics and Civilization* (1934), describes utopic orchestras as synecdoches of the modern age by pairing the scientific calibration of instruments with the division of labor and the sounds of modern industry: "Tempo, rhythm, tone, harmony, melody, polyphony, counterpoint, even dissonance and atonality, were all utilized freely to create a new ideal world, where tragic destiny, the dim longings, the heroic destinies of men could be entertained once more."[31] Sterne advances this scholarship by examining sonic technologies, many of which broadcast or amplify music, but some of which also project voices and even heartbeats.

Phenomenological and psychoanalytical studies tend to focus on the voice, though these occasionally move into music. This scholarship emphasizes epistemologies and heuristics, or, in other words, how we know and how we resolve problems. Don Ihde, for example, argues that thinking itself is an auditory act as we "hear" words in what he calls our "auditory imagination."[32] Walter J. Ong famously argues in *Presence of the Word* that the voice is "the paradigm of all sound for man, sound itself thus of itself suggests presence."[33] Conversely, Jacques Derrida argues that the voice divides presence and falls short of creating or maintaining a unitary selfhood. Instead, the voice produces the illusion of presence.[34] Perhaps this is because it resounds in the body and yet is simultaneously expelled from the body. Psychoanalytical research can be combined with phenomenology through the work of Didier Anzieu and his concept of the "Sonorous Envelope." Anzieu recognizes that experience through senses is the key to psychical growth.[35] Thus, the sonorous envelope is a shield that protects the self and enables it to survive noises that penetrate yet also helps individuals to grow and to thrive. Naturally, these strands of scholarship lead to theorizing about identity formation and selfhood.

Those who follow phenomenological and psychoanalytic theories have recently begun to develop new methods for reading sound. Following Ong, Ihde, and Anzieu, scholars like Steven Feld, Steven Connor, Joshua Gunn, and Mladen Dolar have opened up new avenues for critiquing and understanding the voice. Feld, for example, has been a leading scholar in the study of sonic cultures because he has drawn upon phenomenology. He studies soundscapes and acoustic spaces to understand how environment impacts identity.[36] Connor, a scholar of modern literature who writes about the spoken word and ventriloquism, interprets the voice as a projection outward of the self with the potential to protect the self (as in an "attack" on others) and the power to destroy the self.[37] Like Connor, who seamlessly combines phenomenology with psychoanalysis, Gunn combines both intellectual traditions to hear in recorded voices the division of the self and the possibility of escaping both life and death.[38] Dolar follows up on the discovery by psychologists at the turn of the twentieth century that the sound of the voice exceeds the author's intended meaning.[39] For Dolar (not to mention Connor and Gunn), the voice is a presence, though often a disrupted one that is shattered by its own sound. These scholars have largely limited themselves to the voice and its sound, though their techniques may be extended beyond the body, particularly in light of the fact that sounds are not merely produced by humans, animals, machines, and nature, they are also interpreted.

Scholars trained in humanities' disciplines sometimes draw from Marxian or phenomenological/psychoanalytic theories but also manage to deeply read sound independently of Marx or Sigmund Freud and yet within the traditions of anthropologists, historians, and cultural critics. Philippe-Joseph Salazar, for example, turns to rhetoric and to continental philosophy (ranging from Michel de Certeau to Derrida as well as Ong) in his pathbreaking study *Le Culte de la voix au XVIIe siècle*, where he argues that it is imperative that we listen to the past by tracing the voice through period aesthetics.[40] Literary scholar Wes Folkerth, for example, follows Ihde and Ong in his reading of the sound of William Shakespeare's plays during the seventeenth century. But when Folkerth brilliantly interprets a brief scene that is usually cut from modern performances of Shakespeare's *Antony and Cleopatra*, he weaves sense perception (the core of phenomenology) into a close read of the text. Though we have no ability to hear early performances as these occurred four centuries ago, when no recording equipment existed, Folkerth is still able to describe not only a sound but also the complex meaning of that sound in the context of late sixteenth-century English culture. In *Antony and Cleopatra*, Shakespeare employed shawms, a musical instrument that served as an eerie precursor to the oboe, though the instruments are hidden under the stage during the scene. Shawms, which make buzzing and honking noises like modern-day duck calls, were employed in England during the sixteenth century to announce curfew and daybreak and official occasions like weddings and funerals. Shawm players were also often waits (i.e., town lookouts).[41] Folkerth recognizes a link between Shakespeare's likely intention during the scene and audience expectations about the role of the sound of shawms:

> Owing to the particular duties of the waits, shawms traditionally would have been heard in the dead of the night and in the early morning hours. The sound, coming from outside on the street while one was inside, and often in bed, would have consistently, over generations, accompanied people's transitions from sleep to waking, from unconsciousness to consciousness. The shawm would have been one of the signature sounds of this liminal state of consciousness, and therefore enormously effective at evoking that atmosphere in the theatre.... The sound contributes tonally to the scene, invoking a dramatic sense of the uncanny by activating the early modern audience's extratheatrical experiences with the music of shawms, both ceremonial and nocturnal.[42]

Folkerth's read of the sound of shawms illuminates our understanding of the atmosphere at performances of *Antony and Cleopatra*. More important, Folkerth illustrates how contextualizing sound within its culture enables scholars in the humanities to better understand the society. If Shakespeare employed shawms as a way of drawing upon the audience's sonic experience so that they could be manipulated into witnessing his plays as a medium between historical reality and dream and between the comprehensible and incomprehensible, then we know not only about Elizabethan culture but also that culture shapers have long been employing sound for its persuasive effect. Folkerth is able to understand sound even without the benefit of a sonic recording (though he has listened to shawms). By reading sound, we gain a better understanding of a culture and are better situated to resist the exploitation of sound by modern playwrights, politicians, and corporations.

Folkerth, happily, is not alone. Other scholars have provided insightful analysis of sound that opens up this kind of text in a manner revealing insights into cultures and individuals that have until recently gone unheard and unremarked in humanities scholarship. Like Folkerth, many of these scholars are in English departments. For example, Joel Dinerstein, in his book *Swinging the Machine*, analyzes the sounds of African American music in the 1930s as a response to and incorporation of the mechanization of life. His read of the incorporation of train sounds into jazz music is particularly insightful.[43] While Dinerstein turns sound into words, David Copenhafer turns words into sound. His read of the preface to Ralph Ellison's novel *Invisible Man* brilliantly returns the novelist's thoughts into music, locating for example, bridges (a musical term) that pace the narrative.[44] An older study, David H. Culbert's essay about the voice of Edward R. Murrow, is also equal to the task of restoring sound to words that scholars usually only read from a transcript. Through Culbert's careful ear, Murrow can be heard again pronouncing words in a manner that emphasize moods, impressions, and even events.[45]

To be sure, there are a great many theoretical positions that scholars have developed to better understand sound. Linguists Roman Jakobson and Roland Barthes, for example, theorize the voice and sound independently of the traditions described above, though their theories have proven less pliable and popular.[46] Because the field of sonic studies is still amorphous, the work of theorists from the 1930s also remains viable. Indeed, Rudolf Arnheim's book *Radio* (1936) remains an excellent introduction to theorizing about sound.[47] A. H. Cantril and G. W. Allport's *Psychology of Radio* (1935) is dated and not particularly theoretical and yet is simultaneously insightful in a manner that twenty-first-century scholars are only beginning to rediscover.[48]

Perhaps the most brilliant of all of these radio-era essays was written in 1940 but not published until 2009. The work titled "Radio Physiognomics" combines the Marxian focus on materialism with the phenomenological emphasis on the sensory perception of presence and yet accomplishes something entirely new as well. The essay was written by Adorno for the Princeton Radio Project, a large-scale social-research study of the effects of radio on public perception. In the essay, Adorno argues that the radio voice can be compared to a face because the radio "speaks to us" and, thus, can be examined through the practice of physiognomy. Anticipating Michel Chion's description of the acousmatic voice, Adorno describes the radio voice that emerges from the loud*speaker* as dangerously authoritative. In the presence of such huge and public voices, our own presence becomes tiny, private, and powerless:

> The man who believes that the commentator shouting through his loudspeaker is a virtual dictator is wrong; but the fact that he "sounds like a dictator" is certainly due to conditions which do not allow any voice to be broadcast so that it might fundamentally touch upon the public speaker's illusion of privacy. Thus, in a way, the naïve listener who becomes afraid of the voice of the commentator is right: the social mechanism behind the technical one leading to these disproportions [huge voice and tiny voice] is necessarily one which he has every reason to fear, and it may easily be one which breeds dictators who really shout just as the voice of the humble commentator sounds in a private room.[49]

Obliquely referring to Adolf Hitler and Joseph Goebels, Adorno recognizes the power of the radio to magnify a small voice into an authoritative, godlike voice. We know this voice well. It is the voice of the trickster behind the curtain in Frank L. Baum's *Wizard of Oz*. It is the voice of Lyndon Baines Johnson over the mushroom cloud in the campaign commercial titled "Daisy." It is the voice of conservative radio-host Rush Limbaugh.

We have seen the self both penetrated by sound and protected by sound. And though we might be troubled by the Futurists' embrace of penetrating noises that shattered the self, we should also be concerned about the protective sonic envelope that Adorno recognized in Hitler's sonic manipulations. The penetrating sound is revolutionary and violent. The enveloping sound is reactionary and potentially just as violent. Even as it protects those inside the envelope, it has the potential to assault those outside the envelope. The cheers at Hitler's Nuremburg brought the German people together as an Aryan nation and simultaneously created the motivation and justification for the murders of seven million people not inside that protective envelope.

We have recently seen this dangerous envelope at work in America, though in a less-obvious and less-violent manner. Though affecting liberals (exemplified by Code Pink, antiwar protesters who have spent too much time in an echo chamber that reproduces the same voice repeatedly), I turn toward a more recent and conservative example. During the summer of 2009, as members of the U.S. Congress returned to their states or districts to host town-hall meetings with constituents, anger at the election of a man who had been labeled by voices inside the echo chamber as a foreigner and communist produced a need for sonorous envelopes that protected against otherness and socialist fascism.[50] Prodded in particular by right-wing talk radio, members of the emerging Tea Party, citizens who were certain that the president was born in Kenya, libertarians, and others attended these meetings. But many were only willing to hear their congressperson as long as he or she repeated the same mantras heard on talk radio. When elected officials challenged the voices in the echo chamber, like Republican Representative Bob Inglis of South Carolina, who called on constituents to turn off Glenn Beck because he was a fearmonger, they were shouted down.[51]

These "true" Americans who were concerned that they were losing their country (again, mantras they heard in the echo chamber of right-wing radio) used their voices to protect themselves against the threat of change. Frightened by recently discovered "truths" that America's president was an alien, that the Democrats in Congress and the White House intended to destroy democracy, and that "real" Americans were losing their country to immigrants, gays, and socialists, some people claiming to be patriots showed up at these town-hall meetings to demand proof that the president was born American, to reject any change to the status quo, and to shout down their elected officials. The power of the sonorous envelope is reflected in its ability to persuade others to join in, as Hitler understood, and, thus, attendees who had no intention of being uncivil when they entered the town hall meetings became so, adding to the protective sound while comforting themselves with their own assumptions.

Perhaps the most famous of these encounters was that between Representative Mike Castle, a Delaware Republican, and a woman in a red shirt. It is an existential moment as the woman's very self is threatened with destruction. She begins incoherently, almost forgetting the name of the U.S. Representative she has come to question and then starting a sentence twice and stumbling in a variety of ways over syntax. But she gains strength from the cheers and clapping of the crowd, and her own voice gathers in power as she challenges a member of the U.S. Congress to prove that President Barack Obama is a U.S. citizen. However, emotions catch up to her, particularly as she begins to recount her father's service to the country during World War II, and her syntax is troubled again. But the cheering helps her to put her words (and by extension herself) together. At the end of the diatribe, as if to remember that he is not alone, Castle talks briefly with a supporter who shares the stage with him. Gathering himself, he first admits that he does not know how to respond and then struggles to say the word *president*, repeating the word "*the.*" The normally unflappable Representative is taken aback by the volume of the protest.[52] Notably, Castle avoids the name Obama, referring instead to the office: "the the president there, he is a citizen of the United States."[53]

Now the woman in the red shirt is threatened by the response and calls upon her supporters to surround themselves with the sound of a familiar chant, the Pledge of Allegiance. Though Castle had already called upon someone else to speak, he feels compelled, at the demand of the woman to recite the pledge, to ask if he should lead it. Considering that Castle is a member of the U.S. Congress, stands behind a podium, and has a microphone before him, one might normally think this is a preposterous question. Of course, he should lead the pledge. But he is threatened by the noise of the crowd, and the normally composed representative has been reduced to a puppet in the woman's play. As Castle starts, the crowd does not chant with him in unison, but all come together at the last four words of the pledge's first sentence: "United States of America." After that, Castle is not heard speaking through the microphone (or perhaps at all—his mouth cannot be seen at this point because people have stood up and now block the camera). The audience is cocooned in an envelope of comforting sound that prevents contrary opinions.[54]

Adorno had worried that the power of the radio was not that it created a unified community but rather that it placed listeners, according to John Durham Peters, "into a cocoon of unreflective security or sadistic laughter."[55] This is the product of listening to Limbaugh like he is a friend in the living room or automobile. We are susceptible to manipulation when we trust in our "friends" as we sit with family in the parlor or alone in our cars. That cocoon, mechanically reproduced across a nation of radio commuters who call themselves dittoheads (i.e., people who say "ditto" because they are fans of Limbaugh's show), produces believers who, once gathered, need to hear specific mantras lest their cocoon be torn apart and their selves shattered. Adorno describes a similar phenomenon during the early 1940s, when conformist "Babbitts" who were "profoundly reactionary" aped authority.[56] Once gathered, these Babbitts re-create the cocoon with their own voices by chanting the Pledge of Allegiance or shouting down contrary opinions. This is, of course, a danger to democracy as the sonorous envelope prevents deliberation, argument, and even the acknowledgment of other voices (and, thus, of others). When people believe their selves to be threatened, they act aggressively and loudly to defend themselves. The sonorous envelope repels even as it protects.

Adorno's recently published essay should encourage scholars of media to further question one of the most regularly cited works about popular culture, Walter Benjamin's "The Work of Art in the Age of Mechanical Reproduction," which also dates from this period. Benjamin argues that works of art are made accessible by reproduction, thus destroying a religious aura that prevented original and unique works of art from being analyzed critically. By making art common, Benjamin states, intellectual revolution became possible.[57] Adorno, who had been Benjamin's sponsor, publicly countered the latter's most famous essay in 1944, arguing conversely that sonic reproduction (in particular the radio) creates an aura of authority making dictatorship more likely than revolution and making passive hearing more likely than active listening.[58] Adorno warns of the power of this kind of radio voice.

One last hurdle to the study of sound, and one I have attempted to overcome, is in translation. By translation, I mean a few different though related problems. The first problem is how to translate sounds into words. Ong complains, "The written medium simply will not tolerate all of what actually goes on in oral speech. It has rules. If you cannot fit what you want to verbalize into the rules of writing, you are obligated not to write it."[59] Yet, the critic must attempt to translate sounds into words. The Franklin Delano Roosevelt pause that opens this book is one such translation and a very simple one at that. I can only hope that my readers are able to retranslate my words into an approximation of the original sound. For Benjamin, this is a critical and a metaphorically sonic duty: "The task of the translator consists in finding that intended effect upon the language into which he is translating which produces in it the echo of the original."[60] Unfortunately, my own translations of sound into words are but weak echoes of the originals. Thus, in this volume, citations to readily accessible sonic archives are made wherever possible so that readers may themselves become listeners.

Translation also means that I must translate the visual critical skills I have been taught (both in terms of reading text and reading images) into sonic critical skills. Though Barthes asked and answered, "How, then, does language manage when it has to interpret music? Alas, it seems, very badly," he still equated reading sound to reading words on a page: "Just as the reading of the modern text . . . consists not in receiving, in knowing or in feeling that text, but in writing it anew, in crossing its writing with a fresh inscription, so too reading this Beethoven is to operate his music."[61] In the current volume, I have written anew by reading music and other sounds.

Just as I have been taught to question the meaning of words and images, in listening I can no longer simply accept a sound as a given or as natural; all sounds are interpreted through a culturally imposed framework and in a specific context. This, then, is also a translation of verbal and visual theories into a sonic mode. As a translation, it is necessarily imperfect and subject to revisions and contestation. Indeed, it is my ardent hope that readers will contest my translations. Again, Benjamin employed a sonic metaphor to describe the purpose of the translation: "the language of a translation can—in fact, must—let itself go, so that it gives voice to the *intentio* of the original not as reproduction but as harmony."[62] As such, other translators can recompose sound to create new harmonies and new arguments.

Lastly, translation has the double task of making sense out of very different intellectual traditions that have produced sometimes redundant and sometimes overlapping terms of art. The term *sonorous envelope*, for example, means the sounds that envelop and protect the self (for those who follow psychoanalysis). The same term, under the pen of architect Peg Rawls, applies to spaces, memories, and events through the prism of Henri Bergson's protophenomenology.[63] Sonic-studies scholar Rick Altman uses a similar phrase, *sound envelope*, to describe how edited sounds emerge and fade, and the moments around each sound in a recording.[64] As such, the three definitions of sonic or sound envelope cannot easily be reconciled. Conversely, a different concept receives a variety of different names depending on the intellectual tradition from which the neologism springs: sonic studies, sound studies, acoustemology, and other terms applied to the same field. Similarly, the "period ear" will be found in some books and essays while "aural culture" is in others, and "auditory regime" in still others.[65] Thus, the translator's task in the field of sonic studies is to make sense of often-difficult theoretical work, to find common ground among diverse vocabularies, and to synthesize theories and words into an intelligible *logos*.

I am reminded of Benjamin's cautionary questions: "Will an adequate translator ever be found among the totality of its readers? Or, more pertinently: Does its nature lend itself to translation and, therefore, in view of the significance of the mode, call for it?"[66] I believe the most pertinent question is easily answered: yes. The nature of sound lends itself to translation. Though the vocabulary may be spare, sound is more than significant enough to call critics and scholars to attempt its translation. Moreover, in translating, sound critics will make permanent what is often ephemeral, thus lending sound legitimacy and, to return to Benjamin, a measure of fame. This legitimacy will begin to address the lacuna that scholarship in the humanities has left ever since Plato ruled that music was emotional, opposed to reason, and not worthy of close analysis.

I do not, however, think of translation as a process of reading sound; the two acts are distinct. In the 2008 volume *Listen: A History of Our Ears,* the philosopher Peter Szendy associates translators with musical arrangers: "I always have the impression, in fact, of reading them in the process of reading, of reading their reading of a work. They sign their reading just as arrangers sign their listening. And that is why every reflection on reading—on what it is to read a text—should include the question of translation. That is also why arrangers have so many things to tell us about listening to a work. About what listening means; and about what we can understand by the work. Literally."[67] I no more invent the idea of reading sound than Szendy invents the idea of arrangement or translation. The trick for Szendy and for those scholars who have consciously entered into the study of reading sound is to think about the many ways by which sound has already been and can now be read again. If we are to better understand history and culture and better defend ourselves against the manipulations of the sonically sophisticated, we must learn to read and translate sound.

Translation is one of many sonic products that call on careful critics to think about how sound persuades, effects, and affects. By reading sound can we be made aware of its representations (like historical shifts in the perception of manliness), its benefits (like learning how to adjust to industrial noises), its effects (like the absence of sound denying our ability to empathize), its affects (like the coloring of our perceptions of identity), and its dangers (like the threat the sonorous envelope poses to democracy)? Sound can be read for a variety of elements and in a variety of ways. So I end where I began, with a one-word command from Lawrence W. Levine: listen.

Notes

Preface

I thank Richard Boursy for finding the image of figure 4.1, Benny Goodman and Coleman Hawkins in the studio (1934), in Benny Goodman Papers, Gilmore Music Library, Yale University.

I also thank Don Peterson for permission to republish the photograph in figure 4.2, Jam Session at the Brunswick Studio, March 14, 1937.

1. Michele Hilmes, "Is There a Field Called Sound Culture Studies? And Does It Matter?" *American Quarterly* 57, no. 1 (March 2005): 249.

2. Lawrence W. Levine, "Lawrence Levine Analyzes the Blues," *History Matters,* http://historymatters.gmu.edu/mse/sia/levine.htm, accessed August 2, 2010.

3. Lawrence W. Levine, "William Shakespeare and the American People," in *The Unpredictable Past: Explorations in American Cultural History* (Berkeley: University of California Press, 1993), 163, 338n58.

4. A survey of the *Quarterly Journal of Speech/Quarterly Journal of Speech Education* is illuminating. The word *cadence* was used in four articles during the 2000s, one book review during the 1990s, three articles and a book review in the 1980s, fifteen essays and book reviews in the 1970s, ten essays and book reviews in the 1960s, twenty-one essays and book reviews in the 1950s, twenty essays and book reviews in the 1940s, twenty-two essays and book reviews in the 1930s, and sixteen essays and book reviews in the 1920s.

Chapter 1. Reading Sound

1. Halford R. Ryan, "Roosevelt's First Inaugural: A Study in Technique," *Quarterly Journal of Speech* 65, no. 2 (April 1979): 137–49; Suzanne M. Daughton, "Metaphorical

Transcendence: Images of the Holy War in Franklin Delano Roosevelt's First Inaugural," *Quarterly Journal of Speech* 79, no. 4 (November 1993): 427–46; and Davis W. Houch and Mihaela Nocasian, "FDR's First Inaugural Address: Text, Context, and Reception," *Rhetoric and Public Affairs* 5, no. 4 (2002): 649–78.

2. Lawrence W. Levine and Cornelia R. Levine, *The People and the President: America's Conversation with FDR* (Boston: Beacon, 2002), 16–17.

3. John Erskine, "The Future of Radio as a Cultural Medium," *Annals of the American Academy of Political and Social Science* 177 (January 1935): 214.

4. Quoted in Orrin E. Dunlap Jr., "When 13 Is Lucky," *New York Times*, July 23, 1939.

5. Orrin E. Dunlap Jr., "Hoover, Roosevelt, and Radio," *New York Times*, July 10, 1932.

6. Levine and Levine, *People and the President*, 16; Sherman P. Lawton, "Principles of Effective Radio Speaking," *Quarterly Journal of Speech* 16, no. 3 (June 1930): 270; F. H. Lumley, "Rates of Speech in Radio Speaking," *Quarterly Journal of Speech* 19, no. 3 (June 1933): 403.

7. See Shai Burstyn, "In Quest of the Period Ear," *Early Music* 25, no. 4 (November 1997): 693.

8. Alexander G. Weheliye, *Phonographies: Grooves in Sonic Afro-Modernity* (Durham, NC: Duke University Press, 2005), 10.

9. Plato, interpreted by Joshua Gunn, in Gunn, entry dated July 31, 2007, *Rosewater Chronicles*, http://www.joshiejuice.com/blog/?p=458, accessed July 30, 2010.

10. Leigh Eric Schmidt, *Hearing Things: Religion, Illusion, and the American Enlightenment* (Cambridge, MA: Harvard University Press, 2000), 15–22.

11. Mladen Dolar, *The Voice and Nothing More* (Cambridge, MA: MIT Press, 2006), 127.

12. Marshall McLuhan, "Visual and Acoustical Space," in *Audio Culture: Readings in Modern Music*, ed. Christopher Cox and Daniel Warner (New York: Continuum, 2004), 69.

13. John M. Picker, *Victorian Soundscapes* (New York: Oxford University Press, 2003), 125.

14. William Howland Kenney, *Recorded Music in American Life: The Phonograph in Popular Memory, 1890–1945* (New York: Oxford University Press, 1999), xiii.

15. Michelle Hilmes, *Radio Voices: American Broadcasting, 1922–1952* (Minneapolis: University of Minnesota Press, 1997), xiv.

16. Susan J. Douglas, *Listening In: Radio and the American Imagination* (New York: Times Books, 1999), 7.

17. Penelope Gouk, "In Search of Sound: Authenticity, Healing, and Redemption in the Early Modern State," *Senses & Society* 2, no. 3 (2007): 306.

18. Richard Leppert, "Desire, Power, and the Sonoric Landscape: Early Modernism and the Politics of Musical Privacy," in *The Place of Music*, ed. Andrew Leyshan, David Matless, and George Revill (New York: Guilford, 1998), 292.

19. Lawrence W. Levine, "The Historian and the Culture Gap," in *The Unpredictable Past: Explorations in American Cultural History* (Berkeley: University of California Press, 1993), 14–31.

20. Michael P. Steinberg, *Listening to Reason: Culture, Subjectivity, and Nineteenth-Century Music* (Princeton, NJ: Princeton University Press, 2004), 1.

21. Robert Wess, *Kenneth Burke: Rhetoric, Subjectivity, Postmodernism* (New York: Cambridge University Press, 1996), 39.

22. Kenneth Burke, *Counter-Statement*, 3rd ed. (Berkeley: University of California Press, 1968), 36.

23. Lawrence Grossberg, "Rock, Territoriality, and Power," in *Dancing in Spite of Myself: Essays on Popular Culture* (Durham, NC: Duke University Press, 1997), 99–100; and Stuart Hall, "Calypso Kings," in *The Auditory Culture Reader*, ed. Michael Bull and Les Black (New York: Berg, 2003), 423.

24. Thomas Porcello, "Three Contributions to the 'Sonic Turn,'" *Current Musicology* 83 (Spring 2007): 154.

25. Steven Feld, "From Ethnomusicology to Echo-Muse-Ecology: Reading R. Murray Schafer in the Papua New Guinea Rainforest," *Soundscape Newsletter* 8 (1994): 11; and Mark M. Smith, *Listening to Nineteenth-Century America* (Chapel Hill: University of North Carolina Press, 2001), 262, 318–19n2.

26. Michele Hilmes, "Is There a Field Called Sound Culture Studies? And Does It Matter?" *American Quarterly* 57, no. 1 (March 2005): 249.

27. Jonathan Sterne, *The Audible Past: Cultural Origins of Sound Reproduction* (Durham, NC: Duke University Press, 2004), 3; and R. Murray Schafer, "The Music of the Environment," in *Audio Culture: Readings in Modern Music,* ed. Christopher Cox (New York: Continuum, 2004), 29.

28. Sterne, *Audible Past,* 342–46.

29. Theodor W. Adorno, *Introduction to the Sociology of Music,* trans. E. B. Ashton (1962; New York: Continuum, 1976), 5.

30. Theodor W. Adorno, with George Simpson, "On Popular Music" (1941), in *Essays on Music,* ed. Richard Leppert (Berkeley: University of California Press, 2002), 442–43.

31. Michel Foucault, *The Order of Things: An Archeology of the Human Sciences* (1966; New York: Vintage, 1994), 3–16; Michael Baxandall, *Painting and Experience in Fifteenth-century Italy: A Primer on the Social History of Pictorial Style* (New York: Oxford University Press, 1972), 29–108; and John Berger, *Ways of Seeing* (1972; New York: Viking, 1973).

32. Douglas Kahn, "Art and Sound," in *Hearing History: A Reader,* ed. Mark M. Smith (Athens: University of Georgia Press, 2004), 36.

33. Contents, *Quarterly Journal of Speech* 19, no. 1 (February 1933).

34. A. H. Cantril and G. W. Allport, *The Psychology of Radio* (New York: Harper and Brothers, 1935), 9–10.

35. Ibid., 26, 109, 122–23.

36. Joshua Gunn, "Gimme Some Tongue (On Recovering Speech)," *Quarterly Journal of Speech* 93, no. 3 (2007): 361, 363.

37. Joshua Gunn, "Mourning Speech: Haunting and the Spectral Voices of Nine-Eleven," *Text and Performance Quarterly* 24, no. 2 (2004): 93.

38. Adorno, *Introduction to the Sociology of Music*, 62.

39. Douglas, *Listening In*, 9.

40. Jacques Lacan, "The Mirror Stage as Formative of the Function of the I as Revealed in Psychoanalytic Experience," *Ecrits*, trans. Alan Sheridan (1936; New York: Norton, 1977), 1–7; Michel Foucault, *The Birth of the Clinic: An Archeology of Medical Perception*, trans. A. M. Sheridan Smith (1963; New York: Vintage Press, 1994), 107–23; and Laura Mulvey, "Visual Pleasure and Narrative Cinema," *Screen* 16, no. 3 (Autumn 1975): 6–18.

41. Veit Erlmann, "But What of the Ethnographic Ear? Anthropology, Sound, and the Senses," in *Hearing Cultures*, ed. Erlmann (New York: Berg, 2004), 5.

42. See Fred Newton Scott, *The Standard of American Speech and Other Papers* (Boston: Allyn and Bacon, 1926), 174.

43. Burstyn, "In Quest of the Period Ear," 693.

44. Dunlap, "Hoover, Roosevelt, and Radio."

45. Andrew Thomas Weaver, "Experimental Studies in Vocal Expression," *Quarterly Journal of Speech Education* 10, no. 3 (June 1924): 199.

46. Margarita Alexomanolaki, Catherine Loveday, and Chris Kennet, "Music and Memory in Advertising: Music as a Device of Implicit Learning and Recall," *Music, Sound, and the Moving Image* 1, no. 1 (Spring 2007): 65–68.

47. Kevin Michael DeLuca, Christine Harold, and Kenneth Rufo, "Q.U.I.L.T.: A Patchwork of Reflections," *Rhetoric and Public Affairs* 10, no. 4 (Winter 2007): 640.

48. James Elkins, *The Object Stares Back: On the Nature of Seeing* (New York: Harcourt, 1996).

49. Tom Rice, "Soundselves: An Acoustemology of Sound and Self in the Edinburgh Royal Infirmary," *Anthropology Today* 19, no. 4 (August 2003): 8.

Chapter 2. Fitting Sounds

1. William Howland Kenney, *Recorded Music in American Life: The Phonograph in Popular Memory, 1890–1945* (New York: Oxford University Press, 1999), 24–25.

2. Patrick Feaster and David Giovannoni, "Comstock Laws," in *Actionable Offenses: Indecent Phonograph Recordings from the 1890s,* prod. Feaster and Giovannoni (booklet included with compact disc) (Champaign, IL: Archeophone Records, 2007), 10.

3. Russell Hunting (?), "Out of Order" (1895?), in *Actionable Offenses: Indecent Phonograph Recordings from the 1890s* (booklet included with compact disc) (Champaign, IL: Archeophone Records, 2007), track 8.

4. "An Act for the Suppression of Trade in, and Circulation of, Obscene Literature and Articles of Immoral Use," *Statutes at Large of the United States of America* (1873),

17:598. The phrase "obscene, lascivious, and lewd" is from the Comstock Law, which was enacted in 1873.

5. Richard Bauman and Patrick Feaster, "'Fellow Townsmen and My Noble Constituents!': Representations of Oratory on Early Commercial Recordings," *Oral Tradition* 20, no. 1 (2005): 43.

6. See Susan J. Douglas, *Listening In: Radio and the American Imagination* (New York: Times Books, 1999), 12; and Patrick Feaster and David Giovannoni, "Russell Hunting," in *Actionable Offenses: Indecent Phonograph Recordings from the 1890s* (booklet included with compact disc) (Champaign, IL: Archeophone Records, 2007), 25.

7. See Pierre Bourdieu, *Language & Symbolic Power*, ed. John B. Thompson, trans. Gino Raymond and Matthew Adamson (Cambridge, MA: Harvard University Press, 1991), 45.

8. See Douglas, *Listening In*, 107–20.

9. Jason Camlot, "Early Talking Books: Spoken Recordings and Recitation Anthologies, 1880–1920," in *Book History*, ed. Ezra Greenspan and Jonathan Rose (State College: Penn State University Press, 2003), 6:149.

10. Patrick Feaster and David Giovannoni, "The Swearing Machine," in *Actionable Offenses: Indecent Phonograph Recordings from the 1890s* (booklet included with compact disc) (Champaign, IL: Archeophone Records, 2007), 18.

11. Kenney, *Recorded Music in American Life*, 65.

12. Mary E. Stuckey, *Defining Americans: The Presidency and National Identity* (Lawrence: University Press of Kansas, 2004), 129.

13. Theodore Roosevelt, *The Strenuous Life* (1899; New York: Scribner, 1906), 3.

14. Lawrence W. Levine, "William Shakespeare and the American People," in *The Unpredictable Past: Explorations in American Cultural History* (Berkeley: University of California Press, 1993), 163.

15. Cara Finnegan, "Recognizing Lincoln: Image Vernaculars and Nineteenth-Century Visual Culture," *Rhetoric and Public Affairs* 8, no. 1 (2005): 49–50.

16. Philip S. Foner, *History of the Labor Movement in the United States: The T.U.E.L. to the End of the Gompers Era, 1925–1929*, vol. 10 (New York: International, 1991).

17. Richard Bauman and Patrick Feaster, "Oratorical Footing in a New Medium: Recordings of Presidential Campaign Speeches, 1896–1912," *Texas Linguistic Forum* 47 (2003), http://studentorgs.utexas.edu/salsa/proceedings/2003/bauman&feaster.pdf, accessed August 2, 2010.

18. Oliver Read and Walter L. Welch, *From Tin Foil to Stereo: Evolution of the Phonograph* (1959; New York: Bobbs-Merrill, 1976), 165, 216–17.

19. Ward Marston, "The Gramophone and the Campaigns of 1908 and 1912" (booklet included with compact disc), in *In Their Own Voices: The U.S. Presidential Elections of 1908 and 1912* (Philadelphia: Annenberg School for Communication/Marston Records, 2000), 37–41.

20. Bourdieu, *Language & Symbolic Power*, 45, 53–54, 97.
21. Ibid., 54.
22. Ibid., 68.
23. Kenneth Burke, *A Rhetoric of Motives* (1950; Berkeley: University of California Press, 1969), 19–23.
24. Pierre Bourdieu, *Distinction: A Social Critique of the Judgement of Taste*, trans. Richard Nice (1979; Cambridge, MA: University of Harvard Press, 1984), 255.
25. Bourdieu, *Language & Symbolic Power*, 86–87, 96.
26. George Washington Plunkitt, *Plunkitt of Tammany Hall: A Series of Very Plain Talks on Very Practical Politics,* ed. William L. Riordan (1905; New York: Knopf, 1948), 69.
27. See Charles E. Morris III, "Contextual Twilight/Critical Liminality: J. M. Barrie's Courage at St. Andrews," *Quarterly Journal of Speech* 82, no. 3 (August 1996): 208; and "Pink Herring & the Fourth Persona: J. Edgar Hoover's Sex Crime Panic," *Quarterly Journal of Speech* 88, no. 2 (May 2002): 228.
28. Plunkitt, *Plunkitt of Tammany Hall,* 69–70.
29. Douglas, *Listening In*, 14.
30. See Michael Kimmel, *Manhood in America: A Cultural History* (New York: Free Press, 1996), 120–27.
31. Quoted in Gail Bederman, *Manliness and Civilization: A Cultural History of Gender and Race in America, 1880–1917* (Chicago: University of Chicago Press, 1996), 175.
32. Allen Warren, "Popular Manliness: Baden-Powell Scouting and the Development of Manly Character," in *Manliness and Morality: Middle-Class Masculinity in Britain and America, 1800–1940,* ed. J. A. Mangan and James Walvin (New York: St. Martin's, 1987), 201.
33. John Poole Sandlands, *The Voice and Public Speaking* (London: Hodder and Stoughton, 1880), 104.
34. Dion Boucicault, *The Art of Acting* (New York: Columbia University Press, 1926), 24.
35. Bauman and Feaster, "'Fellow Townsmen and My Noble Constituents!'" 40.
36. Gail Jefferson, "Glossary of Transcript Symbols with an Introduction," in *Conversation Analysis: The First Generation,* ed. G. H. Lerner (Amsterdam: Benjamins, 2004), 13–31.
37. Edward Napoleon Kirby, *Public Speaking and Reading* (Boston: Lee and Shepard, 1896), 104.
38. Grover Cleveland, "Grover Cleveland Gives His 'Front Porch' Campaign Speech," VVL call no. DB25, Vincent Voice Library, Michigan State University Libraries, *Michigan State University, October 11, 2005,* http://vvl.lib.msu.edu/record.cfm?recordid=25, WAV, accessed August 2, 2010.
39. William McKinley, "William McKinley Gives a Campaign Speech from His Front Porch and Talks about the Civil War," VVL call no. DB498, Vincent Voice

Library, Michigan State University Libraries, *Michigan State University,* October 11, 2005, http://grp.lib.msu.edu/vvl.lib.msu.edu/record.cfm?recordid=498, WAV, accessed August 2, 2010.

40. See Bauman and Feaster, "Oratorical Footing in a New Medium," 5, 17.

41. Richard Bebb, "The Voice of Henry Irving: An Investigation," *Recorded Sound* 68 (1977): 730.

42. Gordon Craig, *Henry Irving* (Toronto: Longmans, Green, 1930), 65.

43. David Crystal, *Think on My Words: Exploring Shakespeare's Language* (New York: Cambridge University Press, 2008), 125–30.

44. Jules Zanger, "The Minstrel Show as Theatre of Misrule," *Quarterly Journal of Speech* 60, no. 1 (1974): 35–36.

45. *Oxford English Dictionary,* s.v. "hot air," http://dictionary.oed.com/, accessed July 30, 2010. Mark Twain popularized this colloquialism.

46. L. Frank Baum, "The Wizard of Oz," in *The Wonderful World of Oz* (1900; New York: Penguin, 1998), 79.

47. Rita Reif, "First Half of 'Huck Finn,' in Twain's Hand, Is Found," *New York Times,* February 14, 1991.

48. Mark Twain, *Adventures of Huckleberry Finn* (1884; New York: Norton, 1977), 156.

49. Louis Calvert, *Problems of the Actor* (New York: Holt, 1918), 58.

50. Henry Irving, "Recording of Henry Irving Reciting the Opening Speech of *Richard III,*" *Henry Irving Society,* http://www.theirvingsociety.org.uk/HI-R3.rm, accessed August 2, 2010.

51. Bebb, "Voice of Henry Irving," 727, 730; and Craig, *Henry Irving,* 63–64.

52. Bebb, "Voice of Henry Irving," 731.

53. Calvert, *Problems of the Actor,* 58.

54. Otis Skinner, *Footlights and Spotlights: Recollections of My Life on Stage* (Indianapolis: Bobbs-Merrill, 1924), 350.

55. Brainard Gardner Smith, *Reading and Speaking,* 3rd ed. (Boston: Heath, 1898), 25.

56. Solomon Henry Clark and Frederic Mason Blanchard, *Practical Public Speaking: A Textbook for Colleges and Secondary Schools* (New York: Scribner, 1899), 138, 127.

57. Edwin Du Bois Shurter, *Public Speaking: A Treatise on Delivery* (Boston: Allyn and Bacon, 1903), 26–27.

58. Ibid., 30.

59. Nan Johnson, *Gender and Rhetorical Space in American Life, 1866–1910* (Carbondale: Southern Illinois University Press, 2002), 14, 21, 99.

60. G. Stanley Hall, "The Awkward Age," *Appleton's Magazine,* August 1900, 154.

61. Kimmel, *Manhood in America,* 121.

62. Edward Amherst Ott, "Enemies inside the Elocution Profession," *Werner's Magazine,* November 1901, 203.

63. Ibid., 204.

64. Benjamin Harrison, address to the Pan-American Congress, VVL call no. DB93, Vincent Voice Library, Michigan State University Libraries, *Michigan State University*, http://vvl.lib.msu.edu/showfindingaid.cfm?findaidid=HarrisonB, MP3, accessed August 2, 2010.

65. Garrett A. Hobart, "Recorded for the Opening of the Electrical Exposition of New York City," May 1, 1898, NPS object catalog number EDIS 39849, Edisonia, Documentary Recordings and Political Speeches, National Park Service, *U.S. Department of Interior*, November 5, 2004, http://www.nps.gov/archive/edis/edisonia/documentary.htm, MP3, accessed August 2, 2010.

66. Bebb, "Voice of Henry Irving," 727.

67. George Bernard Shaw, *Pygmalion* (1913; New York: Penguin Books, 2003), 78, 101.

68. Theodore Roosevelt, "The Right of the People to Rule," ca. August 1912, NPS object catalog number EDIS 39850, Edisonia, Documentary Recordings and Political Speeches, National Park Service, *U.S. Department of Interior*, November 5, 2004, http://www.nps.gov/archive/edis/edisonia/documentary.htm, MP3, accessed August 2, 2010.

69. Ibid.

70. B. G. Smith, *Reading and Speaking*, 19.

71. The superscript ° is a Conversation Analysis convention that marks quiet speech.

72. Roosevelt, "Right of the People to Rule."

73. Stuckey, *Defining Americans*, 179.

74. Theodore Roosevelt, *Ranch Life and the Hunting Trail* (1888; New York: Century, 1911), 119.

75. Theodore E. Schmauk, *The Voice, in Speech and Song* (New York: Alden, 1890), 33–34.

76. Ernest Pertwee, *The Art of Speaking* (New York: Putnam's, 1902), 29.

77. David Ffrangcon-Davies, *The Singing of the Future* (Cambridge: Harvard University Press, 1905), 22.

78. Roosevelt, "Right of the People to Rule." Middle C is designated as C_4.

79. Clifford Berryman, "Progressive Fallacies," *Washington Star*, March 18, 1912.

80. Craig H. Roell, *The Piano in America* (Chapel Hill: University of North Carolina Press, 1987), 23–24.

81. Bederman, *Manliness and Civilization*, 170.

82. Kenneth Cmiel, *Democratic Eloquence: The Fight over Popular Speech in Nineteenth-century America* (New York: Morrow, 1990), 250.

83. Kirby, *Public Speaking and Reading*, 100–101.

84. Jeffrey Tullis, *The Rhetorical Presidency* (Princeton: Princeton University Press, 1987), 4; and Ryan L. Teten, "Evolution of the Modern Rhetorical Presidency," *Presidential Studies Quarterly* 33, no. 2 (2003): 339.

85. Elvin T. Lim, "Five Trends in Presidential Rhetoric: An Analysis of Rhetoric from George Washington to Bill Clinton," *Presidential Studies Quarterly* 32, no. 2 (2002): 338–45.

86. Robert J. Higgs, "Yale and the Heroic Ideal: Gotterdammerung and Palingenesis, 1865–1914," in *Manliness and Morality: Middle-Class Masculinity in Britain and America, 1800–1940*, ed. J. A. Mangan and James Walvin (New York: St. Martin's, 1987), 160.

87. William A. Rogers, "Home Again," *Harper's Weekly*, May 13, 1905.

88. Joseph Keppler Jr., "Goodness Gracious! I Must Have Been Dozing!" *Puck*, June 1910.

89. Tullis, *Rhetorical Presidency*; Lim, "Five Trends in Presidential Rhetoric"; and Teten, "Evolution of the Modern Rhetorical Presidency."

90. Levine and Levine, *People and the President*, 15–17.

91. Len Spencer, "The Gettysburg Address," 1903, Antique Audio Show, Community Audio, Audio Archive, *Internet Archive*, January 28, 2008, http://www.archive.org/details/AntiqueAudioShowForJanuary282008, MP3, accessed August 2, 2010.

92. Earl W. Wiley, "Lincoln the Speaker," *Quarterly Journal of Speech* 21, no. 3 (June 1935): 306; and Gary Wills, *Lincoln at Gettysburg: The Words That Remade America* (New York: Simon and Schuster, 1992), 37.

93. William Pettenger, *Extempore Speech: How to Acquire and Practice It* (Philadelphia: Penn, 1899), 119; and B. G. Smith, *Reading and Speaking*, 10.

94. Frederick Houk Law, *Mastery of Speech* (New York: Independent, 1919), 9–10.

95. D. W. Griffiths, dir., *Abraham Lincoln* (Hollywood, CA: United Artists, 1930).

96. Arthur Elliot Sproul, "Roosevelt and Lincoln," *New York Times*, June 20, 1933.

Chapter 3. Machine Mouth

1. Ralph Ellison, "Richard Wright's Blues," in *Living with Music: Ralph Ellison's Jazz Writings*, ed. Robert G. O'Meally (New York: Modern Library, 2002), 101–19.

2. Richard Wright, "King Joe," in *The World of Richard Wright*, ed. Michel Fabre (Jackson: University Press of Mississippi, 2007), 249.

3. Ellison, "Richard Wright's Blues," 113.

4. Gayle Dean Wardlow, *Chasing That Devil Music* (San Francisco: Backbeat, 1998), 103–4.

5. Lewis Mumford, *Technics and Civilization* (1934; New York: Harcourt, Brace, 1963), 14.

6. R. Murray Schafer, *The Soundscape: Our Sonic Environment and the Tuning of the World* (1977; Rochester, VT: Destiny, 1994), 55–56; and Alain Corbin, *Village Bells: The Culture of the Senses in the Nineteenth-century French Countryside* (New York: Columbia University Press, 1998).

7. Mumford, *Technics and Civilization*, 17; and E. P. Thompson, "Time, Work-Discipline, and Industrial Capitalism," *Past and Present* 38 (December 1967): 67.

8. *Oxford English Dictionary Online*, s.vv. "on" (14b), "time" (48a), http://dictionary.oed.com/.

9. Thompson, "Time, Work-Discipline, and Industrial Capitalism," 275–99.
10. Karl Marx, *The Poverty of Philosophy* (Chicago: Kerr, n.d.), 57.
11. John Burroughs, *The Breath of Life* (Boston: Houghton Mifflin, 1915), 95.
12. Walter Benjamin, "On the Concept of History," in *Illuminations* (1940; New York: Schocken, 1969), 261–62; and Benedict Anderson, *Imagined Communities: Reflections on the Origin and Spread of Nationalism* (1983; New York: Verso, 1991), 23–24.
13. Steven Connor, *Dumbstruck: A Cultural History of Ventriloquism* (New York: Oxford University Press, 2000), 16–17.
14. Mumford, *Technics and Civilization*, 197; and Thompson, "Time, Work-Discipline, and Industrial Capitalism," 286.
15. Fritz Lang, director, and Thea von Harbou, writer, *Metropolis* (Babelsburg, Germany: UFA, 1927).
16. Benjamin, "On the Concept of History," 262. The translation is by the current author.
17. Arthur Conan Doyle and John Dixon Carr, "The Adventure of the Seven Clocks," in *The Exploits of Sherlock Holmes* (1954; North Stratford, NH: Ayer, 1999), 25.
18. John McLuskey Jr., "Two Steppin': Richard Wright's Encounter with Blue Jazz," *American Literature* 55 (1983): 332–44.
19. Lawrence Grossberg, *We Gotta Get Out of This Place: Popular Conservatism and Postmodern Culture* (New York: Routledge, 1992), 80–81.
20. Adrian Neculau, "La Societé de communication, une approache psychosociologue," in *Apprende et ensignier dans la societé de communication*, ed. Pierre Chauve et al. (Strasbourg, Germany: Conseil de l'Europe, 2005), 58.
21. Richard Wright, *Native Son* (1940; New York: HarperPerennial, 2005), 3, 35–36, 59, 224.
22. Ibid., 43, 60, 88, 89, 335, 100, 121, 122.
23. Ibid., 165, 253, 255–56, 310.
24. Ibid., 12, 19, 21, 63, 80, 113, 42, 261.
25. Ibid., 58, 92, 150, 215, 15, 222, 334
26. Ibid., 370.
27. Ibid., 3; 165; 80; 217, 262; 304–5; 338, 341, 417, 430.
28. Ibid., 90, 340.
29. Connor, *Dumbstruck*, 13; and Michel Foucault, *Les Anormaux, Cours au Collège de France (1974–75)*, ed. Valerio Marchetti and Antonella Salamoni (Paris: Gallimard, 1999), 187–215.
30. Charles Dickens, "Noises," *All the Year Round* 7, no. 16 (1871): 56–57.
31. Joel Dinerstein, *Swinging the Machine: Modernity, Technology, and African American Culture between the World Wars* (Boston: University of Massachusetts Press, 2003), 44–45.
32. Gilles Deleuze and Felix Guattari, *A Thousand Plateaus: Capitalism and Schizophrenia*, trans. Brian Massumi (1980; Minneapolis: University of Minnesota Press, 1987), 348.

33. Connor, *Dumbstruck,* 21.

34. Herman Melville, "The Paradise of Bachelors and the Tartarus of Maids," in *The Oxford Book of American Short Stories,* ed. Joyce Carol Oates (1855; New York: Oxford University Press, 1992), 82.

35. Rebecca Harding Davis, *Life in the Iron Mills* (1861; Whitefish, MT: Kessinger, 2004), 6.

36. Dickens, "Noises," 57.

37. Thomas Hardy, *Tess of the d'Urbervilles* (1891; New York: Penguin, 2003), 326.

38. Lawrence E. Sullivan, "Sound and Senses: Towards a Hermeneutics of Performance," *History of Religions* 26, no. 1 (1986): 15–16.

39. Hardy, *Tess of the d'Urbervilles,* 327.

40. See Genevieve Lloyd, *Being in Time: Selves and Narrators in Philosophy and Literature* (New York: Routledge, 1993), 96–97; and Norbert Lynton, *The Story of Modern Art* (1980; New York: Phaedon, 1989), 87.

41. Clara Orban, *The Culture of Fragments: Words and Images in Futurism and Surrealism* (Atlanta: Rodopi, 1993), 59–60.

42. Carlo Carra, "The Painting of Sounds, Noises and Smells," in *Manifesto: A Century of Isms,* ed. Mary Ann Caws (1913; Lincoln: University of Nebraska Press, 2001), 201–3.

43. Anonymous, "Futuristen" (1912), quoted in John White, "Futurism and German Expressionism," in *International Futurism in Arts and Literature,* ed. Gunter Berghaus (New York: de Gruyter, 2000), 57.

44. Quoted in ibid., 57.

45. Quoted in Vincent Sherry, *Ezra Pound, Wyndham Lewis, and Radical Modernism* (New York: Oxford University Press, 1993), 13.

46. Paul Wood, *The Challenge of the Avant-Garde* (New Haven: Yale University Press, 1999), 212.

47. *Oxford English Dictionary,* s.vv. "chug," "zipper," "beeping."

48. F. T. Marinetti, "Zong Toomb Toomb," in *Selected Poems and Related Prose,* ed. Luce Marinetti, trans. Elizabeth R. Napier and Barbara R. Studholme (1914; New Haven: Yale University Press, 2002), 75.

49. F. T. Marinetti, "Destruction of Syntax—Imagination without Strings—Words-in-Freedom," in *Modernism: An Anthology,* ed. Lawrence Rainey (1913; New York: Blackwell, 2005), 32.

50. Orban, *Culture of Fragments,* 36.

51. F. T. Marinetti, "La Battaglia di Adrianopoli," 1924, Pennsound: Center for Programs in Contemporary Writing, *University of Pennsylvania,* http://writing.upenn.edu/pennsound/x/Marinetti.html, MP3.

52. Orban, *Culture of Fragments,* 36.

53. Kenneth Burke, "The Rhetoric of Hitler's Battle," in *Philosophy of Literary Form: Studies in Symbolic Action* (1939; Baton Rouge: Louisiana State University Press, 1941), 200, 213n6.

54. Quoted in Jacques Attali, *Noise: The Political Economy of Music* (1977; Minneapolis: University of Minnesota Press, 1992), 87.

55. Stephen Connor, "Violence, Ventriloquism, and the Vocalic Body," in *Psychoanalysis and Performance*, ed. Patrick Campbell and Adrian Kear (New York: Routledge, 2001), 81.

56. Grover Cleveland, "Grover Cleveland Gives His 'Front Porch' Campaign Speech," VVL call no. DB25, Vincent Voice Library, Michigan State University Libraries, *Michigan State University*, http://vvl.lib.msu.edu/record.cfm?recordid=25, accessed August 2, 2010; and Russell Hunting, "Out of Order," *Actionable Offenses: Indecent Phonograph Recordings from the 1890s* (Champaign, IL: Archeophone Records, 2007), track 8.

57. A. H. Cantril and G. W. Allport, *The Psychology of Radio* (New York: Harper and Brothers, 1935), 260.

58. Steven Connor, "The Help of Your Good Hands: Reports on Clapping," in *The Auditory Culture Reader*, ed. Michael Bull and Les Back (New York: Berg, 2003), 68, 69.

59. Quoted in Ulrich Schönherr, "Topophony of Fascism: On Marcel Beyer's *The Karneau Tapes*," *Germanic Review* 73 (1998): 329.

60. Theodor W. Adorno, "Radio Physiognomics," in *Current of Music: Elements of a Radio Theory*, ed. Robert Hullot-Kentor (2006; Malden, MA: Polity, 2009), 70–72.

61. Quoted in Schönherr, "Topophony of Fascism," 338.

62. Mary Devereaux, "Beauty and Evil: The Case of Leni Riefenstahl's *Triumph of the Will*," in *Aesthetics and Ethics: Essays at the Intersection*, ed. Jarrold Levinson (New York: Cambridge University Press, 1998), 228–29.

63. Charlie Chaplin, writer, director, *The Great Dictator* (Hollywood, CA: United Artists, 1939).

64. Lynton, *Story of Modern Art*, 151.

65. Quoted in Samantha Rippner, "Urban Life/Urban Dynamism," in *Rhythms of Modern Life: British Prints, 1914–1939*, ed. Clifford S. Ackley (Boston: Museum of Fine Art, 2008), 94.

66. Quoted in Shannon Mattern and Barry Salmon, "Sound Studies: Framing Noise," *Music, Sound and the Moving Image* 2, no. 2 (Autumn 2008): 139.

67. Quoted in ibid.

68. Jesse Bradley, "Hammer Ring," *Negro Work Songs and Calls* (Rounder, 1999), track 13, CD.

69. Alex Ross, *The Rest Is Noise: Listening to the Twentieth Century* (New York: Farrar, Straus, and Giroux, 2007), 98, 180; and Lynton, *Story of Modern Art*, 151.

70. Frank Moore Colby, Talcott Williams, and Herbert Treadwell Wade, *The New International Encyclopedia* (New York: Dodd, Meade, 1922), s.v. "sound transmitters," 645.

71. Laurence Marsham Cockaday, *Radio-Telephony for Everyone* (New York: Stokes, 1922), 25–26.

72. Didier Anzieu, "L'Enveloppe sonore du soi," *Nouvelle revue de pyschanalyse* 13 (Spring 1976): 167–70.
73. Brandon LaBelle, *Background Noise: Perspectives in Sound Art* (New York: Continuum, 2006), 130.
74. Connor, *Dumbstruck*, 29.
75. Didier Anzieu, *The Skin Ego*, trans. Chris Turner (New Haven: Yale University Press, 1989), 169.
76. Elaine Scarry, *The Body in Pain: The Making and Unmaking of the World* (New York: Oxford University Press, 1985), 6.
77. Deleuze and Guattari, *Thousand Plateaus*, 310.
78. Gene Richards, "On the Assembly Line," *Atlantic Monthly*, April 1937, 428.
79. D. H. Lawrence, *The Rainbow* (1915; New York, 1943), 6.
80. Schafer, *Soundscape*, 10, 152.
81. Thomas Mann, "A Man and His Dog," in *Stories of Three Decades* (1918; New York: Knopf, 1936), 440–41.
82. Paul Deharme, *Pour un art radiophonique* (Paris: Le Rouge et Le Noir, 1930), 29.
83. Edith Lecourt, "The Musical Envelope," in *Psychic Envelopes*, ed. Didier Anzieu, trans. Daphne Briggs (London: Karnac Books, 1990), 227–28.
84. Theodor W. Adorno, "The Radio Symphony," in *Essays on Music*, ed. Richard Leppert, trans. Susan H. Gillespie (1941; Berkeley: University of California Press, 2002), 256.
85. Walt Whitman, *Leaves of Grass* (1855; Radford, VA: Wilder, 2007), 327.
86. Norm Cohen, *Long Steel Rail: The Railroad in American Folksong* (1981; Chicago: University of Chicago Press, 2000), 649.
87. Schafer, *Soundscape*, 81.
88. Patrick Feaster and David Giovannoni, "Russell Hunting," in *Actionable Offenses: Indecent Phonograph Recordings from the 1890s* (booklet included with compact disc) (Champaign, IL: Archeophone Records, 2007), 26; Friedrich A. Kittler, *Gramophone, Film, Typewriter*, trans. Geoffrey Winthrop-Young and Michael Wentz (1986; Stanford, CA: Stanford University Press, 1999), 37; and Cohen, *Long Steel Rail*, 442.
89. Dinerstein, *Swinging the Machine*, 98.
90. Michael Jarrett, "Train Tracks: How the Railroad Rerouted Our Ears," *Strategies* 14 (2001): 32.
91. Whitman, *Leaves of Grass*, 327.
92. Dinerstein, *Swinging the Machine*, 117.
93. Cockaday, *Radio-Telephony for Everyone*, 25–26.
94. Quoted in Kittler, *Gramophone, Film, Typewriter*, 71.
95. Adorno, "Radio Symphony," 258.
96. Quoted in Timothy D. Taylor, "Music and the Rise of Radio in Twenties Amer-

ica: Technological Imperialism, Socialization, and the Transformation of Intimacy," in *Wired for Sound: Engineering and Technologies in Sonic Cultures,* ed. Paul D. Greene and Thomas Porcello (Middletown, CT: Wesleyan University Press, 2005), 255.

Chapter 4. The Race of Sound

1. Nathan Porterfield, *Jimmie Rodgers: The Life and Times of America's Blue Yodeler* (Urbana: University of Illinois Press, 1992), 258–60; Episode 38, *Johnny Cash Show,* American Broadcasting Company, October 28, 1970; and Jimmie Rodgers, "Blue Yodel #9," *RCA: Country Legends,* 1930, RCA/BMG Heritage, 2002, CD, track 6.

2. Jules Tygiel, *Baseball's Great Experiment: Jackie Robinson and His Legacy* (1983; New York: Oxford University Press, 2008).

3. Saidiya V. Hartman, *Scenes of Subjection: Terror, Slavery, and Self-making in Nineteenth-century America* (New York: Oxford University Press, 1997); Cheryl I. Harris, "Whiteness as Property," *Harvard Law Review* 106 (1993): 1709–95; and Shane White and Graham White, *Stylin': African American Expressive Culture from Its Beginnings to the Zoot Suit* (Ithaca: Cornell University Press, 1999).

4. See Winthrop D. Jordan, *White Over Black: American Attitudes toward the Negro, 1550–1812* (Chapel Hill: University of North Carolina Press, 1968). See also C. Vann Woodward, *The Strange Career of Jim Crow* (1955; New York: Oxford University Press, 1974); and Ronald Takaki, *Iron Cages: Race and Culture in Nineteenth-century America* (1979; New York: Oxford University Press, 1990).

5. Lawrence W. Levine, "Slave Songs and Slave Consciousness: Explorations in Neglected Sources," in *The Unpredictable Past: Explorations in American Cultural History* (New York: Oxford University Press, 1993), 36–58.

6. See David Copenhafer, "Invisible Music (Ellison)," in *Sonic Interventions,* ed. Sylvia Mieszkowski, Joy Smith, and Marijke de Valck (New York: Rodopi, 2007), 172.

7. Burton W. Peretti, *The Creation of Jazz: Music, Race and Culture in Urban America* (Urbana: University of Illinois Press, 1992); and Kathy A. Ogren, *The Jazz Revolution* (New York: Oxford University Press, 1989).

8. Rick Kennedy, *Jelly Roll, Bix, and Hoagy: Gennett Studios and the Birth of Recorded Jazz* (Bloomington: Indiana University Press, 1999), 36, 148; and Nancy K. McLean, *Behind the Mask of Chivalry: The Making of the Second Ku Klux Klan* (New York: Oxford University Press, 1995), 18.

9. Kennedy, *Jelly Roll, Bix, and Hoagy,* 14–15, 18–25.

10. Alyn Shipton, *New History of Jazz* (New York: Continuum, 2001), 108, 77.

11. Woodward, *Strange Career of Jim Crow,* 6–7.

12. New Orleans Rhythm Kings and Jelly Roll Morton, *New Orleans Rhythm Kings and Jelly Roll Morton,* Milestone 1991, CD.

13. Teddy Wilson, with Arie Ligthart and Humphrey van Loo, *Teddy Wilson Talks Jazz* (New York: Cassell, 1996), 32.

14. Jimmie Rodgers, "Let Me Be Your Sidetrack," *RCA Country Legends*, 1930, RCA/BMG Heritage 2002, CD, track 7.

15. Charles Creaths's Jazz-o-Maniacs, "Grandpa's Spell," *Jazz in Saint Louis, 1924–27*, 1927, CTI 1997, CD, track 14; and Amédé Ardoin, "La Valse à Austin Ardoin," *I'm Never Comin' Back*, 1929; Arhoolie, 1995, CD, track 2.

16. Carl Engel, "Jazz: A Musical Discussion," in *119 Years of the Atlantic*, ed. Louise Desaulniers (1922; New York: Atlantic Monthly, 1977), 294.

17. Shipton, *New History of Jazz*, 237, 322; Ogren, *Jazz Revolution*, 160–61; and Peretti, *Creation of Jazz*, 185.

18. James Lincoln Collier, *Benny Goodman and the Swing Era* (New York: Oxford University Press, 1989), 120–21.

19. Peretti, *Creation of Jazz*, 200.

20. Kennedy, *Jelly Roll, Bix, and Hoagy*, 52.

21. Wilson, *Teddy Wilson Talks Jazz*, 32.

22. John Durham Peters, *Speaking into Air: A History of the Idea of Communication* (Chicago: University of Chicago Press, 1999), 197–98.

23. Richard Bauman and Patrick Feaster, "'Fellow Townsmen and My Noble Constituents!': Representations of Oratory on Early Commercial Recordings," *Oral Tradition* 20, no. 1 (2005): 37.

24. John M. Picker, *Victorian Soundscapes* (New York: Oxford University Press, 2003), 123.

25. Peters, *Speaking into Air*, 211.

26. Susan J. Douglas, *Listening In: Radio and the American Imagination* (New York: Times Books, 1999), 41–42.

27. Bram Stoker, *Dracula* (1897; New York: Oxford University Press, 1996), 221, 335.

28. L. Frank Baum, *The Patchwork Girl of Oz* (1913; Chicago: Reilly and Lee, 1936), 89.

29. Arthur Conan Doyle, "The Adventure of the Mazarin Stone," in *Sherlock Holmes: The Complete Novels and Stories* (1927; New York: Bantam, 1986), 2:570–73.

30. Theodor W. Adorno, "Opera and the Long-playing Record," in *Essays on Music*, ed. Richard Leppert, trans. Thomas Y. Levin (Berkeley: University of California Press, 2002), 284.

31. Rudolph Arnheim, *Radio*, trans. Margaret Ludwig and Herbert Read (1936; London: Faber and Faber, n.d.), 145.

32. Igor Stravinsky, *Igor Stravinsky: An Autobiography* (1936; New York: Norton, 1962), 72.

33. John Erskine, "The Future of Radio as a Cultural Medium," *Annals of the American Academy of Political and Social Science* 177 (January 1935): 217.

34. Ralph Ellison, *Invisible Man* (1952; New York: Vintage Books, 1980), 8.

35. Laszlo Moholy-Nagy, "Production-Reproduction," in *Photography in the Mod-*

ern Era: European Documents and Critical Writings, 1913–1940, ed. Christopher Phillips (New York: Metropolitan Museum of Art, 1989), 81.

36. Joseph Witek, "Blindness as a Rhetorical Trope in Blues Discourse," *Black Music Research Journal* 8, no. 2 (Autumn 1988): 177–78.

37. Edward Abbe Niles, "Ballads, Songs and Snatches," *Bookman* 68, no. 3 (November 1928): 328.

38. Charles L. Black Jr., "My World with Louis Armstrong," *Yale Review* 69, no. 1 (Autumn 1979): 2.

39. Peretti, *Creation of Jazz*, 199, 208, 191; and Artie Shaw, *The Trouble with Cinderella* (New York: Farrar, Straus, and Young, 1952), 229.

40. Max Jones and John Chilton, *Louis: The Louis Armstrong Story, 1900–1971* (Boston: Little, Brown, 1971), 113, 115.

41. Carl Van Vechten, *Nigger Heaven* (Urbana: University of Illinois Press, 2000), 12.

42. Langston Hughes, "Harlem Night Club," in *The Weary Blues* (New York: Knopf, 1926), 32.

43. Ralph Ellison, "Ralph Ellison's Territorial Vantage," in *Living with Music*, ed. Robert G. O'Malley (New York: Modern Library, 2001), 28.

44. Eric Kilgerman, *Sites of the Uncanny: Paul Celan, Specularity and the Visual Arts* (New York: Walter de Gruyter, 2007), 2–3.

45. Collier, *Benny Goodman and the Swing Era*, 120.

46. Publicity still, in Busby Berkeley, *Hollywood Hotel,* Hollywood: Warner Brothers, 1937.

47. Gary Wills, *Lincoln at Gettysburg: The Words That Remade America* (New York: Simon and Schuster, 1992), 36.

48. Jordan, *White Over Black*, xi, 252–59.

49. Steven Connor, *Dumbstruck: A Cultural History of Ventriloquism* (New York: Oxford University Press, 2000), 4.

50. Walter J. Ong, *The Presence of the Word: Some Prolegomena for Cultural and Religious History* (New Haven, CT: Yale University Press, 1967), 112.

51. Billie Holiday and William Dufty, *Lady Sings the Blues* (New York: Lancer Books, 1965), 41.

52. Benny Goodman and His Orchestra, "Your Mother's Son-in-Law," *Lady Day's Greatest, 1933–1944,* ASV 1996, CD, track 1; Donald Clarke, *Wishing on the Moon: The Life and Times of Billie Holiday* (New York: Viking, 1994), 74, 174–75; and Holiday and Dufty, *Lady Sings the Blues,* 32, 34, 37.

53. Benny Goodman's Orchestra, "Bei Mir Bist du Schoen," *Benny Goodman at Carnegie Hall, 1938,* Columbia 1999, CD, track 2:6.

54. Artie Shaw, "Any Old Time," *The Very Best of Artie Shaw,* RCA/Victor, 2001, CD, track 1.

55. Holiday and Dufty, *Lady Sings the Blues,* 75.

56. "Charles Peterson Goes to a Party (1939)," *Jazz Lives*, December 12, 2009, http://jazzlives.wordpress.com/2009/12/12/charles-peterson-goes-to-a-party-1939/.

57. Emma Grant Meader, "Sound and Rhythm in the Speech of Children," *Quarterly Journal of Speech* 20, no. 2 (April 1934): 278.

58. Anne Shaw Faulkner, "Does Jazz Put the Sin in Syncopation?" *Ladies Home Journal*, August 1921, 16.

59. Sophie A. Pray, "Phonetics Training for Children—a Function of the Normal School," *Quarterly Journal of Speech Education* 9, no. 2 (April 1923): 163.

60. Peretti, *Creation of Jazz*, 84.

61. Pierre Bourdieu, *Language & Symbolic Power*, ed. John B. Thompson, trans. Gino Raymond and Matthew Adamson (Cambridge, MA: Harvard University Press, 1991), 45–54.

62. Joshua Gunn, "The Audio Archive Is the Mother of Memorials," September 23, 2007, *Rosewater Chronicles*, http://www.joshiejuice.com/blog/?p=482.

63. Leigh Eric Schmidt, *Hearing Things: Religion, Illusion, and the American Enlightenment* (Cambridge, MA: Harvard University Press, 2000), 170.

64. Andrew Thomas Weaver, "Experimental Studies in Vocal Expression," *Quarterly Journal of Speech Education* 10, no. 3 (1924): 199.

65. Ong, *Presence of the Word*, 140.

66. Ibid., 130.

67. "Interview with Aunt Phoebe Boyd, Dunnsville, Virginia, 1935," 8 parts with transcription, "Voices from the Days of Slavery," *American Memory: Library of Congress*, http://memory.loc.gov/cgi-bin (accessed July 20, 2010).

68. Mark Katz, *Capturing Sound: How Technology Has Changed Music* (Berkeley: University of California Press, 2004), 15.

69. Lee Cataldi, review of *Classical Rhetoric in English Poetry* by Brian Vickers, *Modern Language Review* 66, no. 2 (1971): 381.

70. Don H. Bialostosky and Lawrence D. Needham, *Rhetorical Traditions and British Romantic Literature* (Bloomington: Indiana University Press, 1995), 53.

71. Henry David Thoreau, entry, December 31, 1837, in *The Heart of Thoreau's Journals*, ed. Odell Shepherd (New York: Courier Dover, 1961), 3.

72. A. H. Cantril and G. W. Allport, *The Psychology of Radio* (New York: Harper and Brothers, 1935), 109.

73. T. J. H. McCarthy, "Literary Practice in Eleventh-century Music Theory: The *Colores Rhetorici* and Aribo's *de Musica*," *Medium Aevum* 71, no. 2 (September 2002): 191–92.

74. See J. W. Shoemaker, *Advanced Elocution: Designed as a Practical Treatise for Teachers and Students in Vocal Training, Articulation, Physical Culture, and Gesture* (1898; Philadelphia: Penn, 1927), 84; David Ffrangcon-Davies, *Singing of the Future* (London: John Lane, 1905), 118; Agnes C. Loughlin, "The Voice in Speaking and Singing," *Quarterly Journal of Speech Education* 6, no. 2 (April 1920): 15; and Floyd K.

Riley, "The Conversational Basis of Public Address," *Quarterly Journal of Speech* 14, no. 4 (November 1928): 237.

75. *Comprehensive Standard Dictionary of the English Language* (New York: Funk & Wagnalls, 1933), s.v. "color."

76. Lawrence W. Levine and Cornelia R. Levine, *The People and the President: America's Conversation with FDR* (Boston: Beacon, 2002), xi; and Elena Razlogova, "True Crime Radio and Listener Disenfranchisement with Network Broadcasting, 1935–1946," *American Quarterly* 58, no. 1 (March 2006): 140.

77. Douglas, *Listening In*, 85, 18; and Michelle Hilmes, *Radio Voices: American Broadcasting, 1922–1952* (Minneapolis: University of Minnesota Press, 1997), xix.

78. George J. Buelow, "Rhetoric and Music," in *New Grove Dictionary of Music and Musicians*, ed. Stanley Sadie (London: Macmillan, 1980), 15:793–95.

79. Michael Denning, *The Cultural Front: The Laboring of American Culture in the Twentieth Century* (New York: Verso, 1998), 323–62.

80. David W. Stowe, "The Politics of Café Society," *Journal of American History* 84, no. 4 (March 1998): 1388–91; and Ellison, "Ralph Ellison's Territorial Vantage," 27.

81. Billie Holiday, "Strange Fruit," *Lady Day's 25 Greatest, 1933–1944*, ASV, 1996, CD, track 25.

82. Sigmund Freud, *The Uncanny*, trans. David Mclintock (1919; New York: Penguin, 2003), 132–33, 138.

83. Joshua Gunn, "On Answering Machines and the Voice Abject," *Liminalities* 3, no. 1 (2007), http://liminalities.net/3-1/machine/machine.htm.

84. Mladen Dolar, *The Voice and Nothing More* (Cambridge, MA: MIT Press, 2006), 7–8.

85. Eric Kligerman, *Sites of the Uncanny: Paul Celan, Specularity, and the Visual Arts* (New York: de Groyter, 2007), 56.

86. Rudolph Lothar, *The Talking Machine: A Technical-Aesthetic Essay* (Leipzig: N.p., 1924), 58.

87. Frank Swinnerton, "A Defence of the Gramaphone," *Gramaphone* 1, no. 1 (1923): 52–53.

88. Schmidt, *Hearing Things*, 170; and Peters, "The Uncanniness of Mass Communication," 111.

89. Lisa Gitelman, *Scripts, Grooves, and Writing Machines: Representing Technology in the Edison Era* (Stanford: Stanford University Press, 1999), 122.

90. William Howland Kenney, *Recorded Music in American Life: The Phonograph in Popular Memory, 1890–1945* (New York: Oxford University Press, 1999), 39.

91. Walter B. Gibson and Orson Welles, "The Silent Avenger," *The Shadow*, Mutual Broadcasting System, March 13, 1938.

92. Ibid., "Deathhouse Rescue," *The Shadow*, Mutual Broadcasting System, September 26, 1937.

93. Mel Watkins, *On the Real Side: A History of African American Comedy* (New York: Simon and Schuster, 1994), 568–70.

94. Barbara Dianne Savage, *Broadcasting Freedom: Radio, War, and the Politics of Race, 1938–1948* (Chapel Hill: University of North Carolina Press, 1999), 6–7.
95. Douglas, *Listening In,* 107.
96. "Amos and [sic] Andy," *Old Time Radio Network Library,* http://www.otr.net/?p=amnd.
97. See Douglas, *Listening In,* 108.
98. Joseph B. Treaster, "Freeman F. Gosden Is Dead at 83; Amos in Radio's Amos 'n' Andy," *New York Times,* December 11, 1982.
99. Douglas Kahn, *Noise Water Meat: A History of Sound in the Arts* (Cambridge, MA: MIT Press, 1999), 47.
100. Russel Stamm, *The Invisible Scarlet O'Neil* (1941–43; Whitefish, MT: Kessinger, 2005), 134.
101. F. Scott Fitzgerald, "Dice, Brassknuckles, & Guitar," in *Tales of the Jazz Age,* ed. James L. W. West (New York: Cambridge University Press, 2002), 287–88.
102. Bourdieu, *Language and Symbolic Power,* 95–96.

Chapter 5. Sounds of War

1. Langston Hughes, "World without End," in *The Collected Works of Langston Hughes,* ed. Joseph McLaren (1937; Columbia: University of Missouri Press, 2001), 323.
2. David H. Culbert, "This Is London: Edward R. Murrow, Radio News, and American Aid to Britain," *Journal of Popular Culture* 10, no. 1 (Summer 1976): 33.
3. Kenneth Burke, *A Rhetoric of Motives* (1950; Berkeley: University of California Press, 1969), 19–27.
4. Edward R. Murrow, "Radio Days: London after Dark," *James F. Widner,* 2004, http://www.otr.com/londonafterdark.shtml (accessed August 1, 2010).
5. Philip Seib, *Broadcasts from the Blitz: How Edward R. Murrow Helped Lead America into War* (Washington, DC: Potomac Books, 2006), 3.
6. Culbert, "This Is London," 29–30.
7. Elwyn A. Mauck, "History of Civil Defense in the United States," *Bulletin of the Atomic Scientists* 6 (August–September 1950): 266; and Allan M. Winkler, *Life under a Cloud: American Anxiety about the Atom* (Urbana: University of Illinois Press, 1999), 110.
8. Christopher Simpson, *Science of Coercion: Communication Research and Psychological Warfare, 1945–1960* (New York: Oxford University Press, 1994), 55.
9. Ibid., 55.
10. Timothy Richard Glander, *Origins of Mass Communications Research during the American Cold War: Educational Effects and Contemporary Implications* (Mahwah, NJ: Erlbaum, 2000), 134–36.
11. Guy Oakes, *The Imaginary War: Civil Defense and the American Cold War Culture* (New York: Oxford University Press, 1994), 47.

12. "Fear of Atom Raid Voiced in Report," *New York Times,* January 3, 1953.

13. Harry S. Truman, "Executive Order 10186—Establishing the Federal Civil Defense Administration in the Office for Emergency Management of the Executive Office of the President," December 1, 1950, American Presidency Project, University of California at Santa Barbara, http://www.presidency.ucsb.edu/ws/index.php?pid=78352.

14. Laura McEnaney, *Civil Defense Begins at Home: Militarization Meets Everyday Life in the Fifties* (Princeton: Princeton University Press, 2000), 12.

15. Oakes, *Imaginary War,* 57, 50–51.

16. McEnaney, *Civil Defense Begins at Home,* 36.

17. Dee Garrison, *Bracing for Armageddon: Why Civil Defense Never Worked* (New York: Oxford University Press, 2006), 47.

18. Ishiro Honda, writer and director, *Godzilla* (Tokyo: Toho, 1954); and Bill Haley and His Comets, "Thirteen Women (and Only One Man in Town)," (Decca, 1954), 33.

19. R. Murray Schafer, *The Soundscape: Our Sonic Environment and the Tuning of the World* (1977; Rochester, VT: Destiny, 1994), 272.

20. Michel Chion, *Audio-Vision: Sound on Screen,* trans. Claudia Gorbman (1990; New York: Columbia University Press, 1994), 5.

21. Ibid., 121.

22. Daniel Dervin, "The Primal Scene and the Technology of Perception in Theater and Film," *Psychoanalytic Review* 62, no. 2 (Summer 1975): 290.

23. Marshall McLuhan, *Understanding Media: The Extensions of Man* (1964; New York: Routledge, 2001), 33–34.

24. J. Michael Hogan, "Science of Cold War Strategy: Propaganda and Public Opinion in the Eisenhower Administration's 'War of Words,'" in *Critical Reflections on the Cold War,* ed. Martin Medhurst and H. W. Brands (College Station: Texas A&M University Press, 2000), 157.

25. McEnaney, *Civil Defense Begins at Home,* 50.

26. Robert Hariman and John Louis Lucaites, *No Caption Needed: Iconic Photographs, Public Culture and Liberal Democracy* (Chicago: University of Chicago Press, 2007), 7–12.

27. Stephen Connor, "Sound and Self," in *Hearing History: A Reader,* ed. Mark Smith (Athens: University of Georgia Press, 2004), 59.

28. *Fast and Furry-ous,* dir. Chuck Jones (Hollywood: Warner Brothers, 1949).

29. Douglas Kahn, *Noise Water Meat: A History of Sound in the Arts* (Cambridge, MA: MIT Press, 1999), 11.

30. Chuck Jones, *Conversations,* ed. Maureen Furniss (Jackson: University Press of Mississippi, 2005), 55; and Chuck Jones, *Chuck Amuck: The Life and Times of an Animated Cartoonist* (New York: Farrar, Straus, and Giroux, 1989), 226.

31. Jones, *Conversations,* 58.

32. Jones, *Chuck Amuck,* 219.

33. Ibid, 220–21.

34. Tom Sito, *Drawing the Line: The Untold Story of the Animation Unions from Bosko to Bart Simpson* (Lexington: University Press of Kentucky, 2006), 4–5, 23, 128, 377–78.

35. Michael E. Birdwell, "'Oh, You Thing from Another World, You': How Warner Bros. Animators Responded to the Cold War (1948–1980)," *Film and History* 31, no. 1 (May 2001): 34–35.

36. Roger Chapman, "George F. Kennan as Represented by Chuck Jones: *Road Runner* and the Cold War Policy of Containment (1949–1980)," *Film and History* 31, no. 1 (May 2001): 40–42.

37. Margalit Fox, "Tony Schwartz, Father of 'Daisy Ad' for the Johnson Campaign, Dies at 84," *New York Times*, June 17, 2008.

38. Chion, *Audio-Vision*, 34.

39. McLuhan, *Understanding Media*, 84.

40. Michel Chion, *The Voice in Cinema*, trans. Claudia Gorbineau (1982; New York: Columbia University Press, 1999), 17–29; and Chion, *Audio-Vision*, 71–73, 128–31.

41. Chion, *Voice in Cinema*, 24; Chion, *Audio-Vision*, 72.

42. Tony Schwartz, *Media, the Second God* (Garden City, NY: Anchor Books, 1983), 4.

43. Russell Freeburg, "G.O.P. Pledges Sane TV Ads for Campaign: Rips Johnson Aids [sic] for Horror Commercial," *Chicago Tribune*, September 14, 1964.

44. Freeburg, "G.O.P. Pledges Sane TV Ads for Campaign."

45. Scott Jacobs, "Nonfallacious Rhetorical Strategies: Lyndon Johnson's Daisy Ad," *Argumentation* 20, no. 4 (December 2006): 432.

46. John D. Morris, "Parties Sign Fair-Play Pledge, Then Wrangle over Johnson Ad," *New York Times*, September 12, 1964.

47. Jacobs, "Nonfallacious Rhetorical Strategies," 433.

48. Charles Mohr, "Goldwater Lays War Aim to Reds: He Says No U.S. President Would Start Any Conflict," *New York Times*, October 2, 1964.

49. Fox, "Tony Schwartz, Father of 'Daisy Ad.'"

50. Tony Schwartz, *The Responsive Chord: How Radio and TV Manipulate You . . . Who You Vote for . . . What You Buy . . . and How You Think* (1969; Garden City, NY: Anchor Books, 1974), 93.

51. Ibid., 27–28.

52. Kathleen Hall Jamieson, interview by David Hoffman, "Prof. Kathleen Hall Jamieson on Tony Schwartz," *Tony Schwartz: Media Pioneer, Audio Documentarian*, http://tonyschwartz.org/JamiesonInterview.html, accessed August 1, 2010.

53. Schwartz, *Media, the Second God*, 46.

54. Alfred McCoy, *A Question of Torture: CIA Interrogation from the Cold War to the War on Terror* (New York: Holt, 2006), 24.

55. Michael Otterman, *American Torture: From the Cold War to Abu Ghraib and Beyond* (Carlton, Australia: University of Melbourne Press, 2007), 100.

56. Suzanne G. Cusick, "'You are in a place that is out of the world . . .': Music in the Detention Camps of the 'Global War on Terror,'" *Journal of the Society of American Music* 2, no. 1 (2008): 3–4.

57. Steve Goodman, *Sonic Warfare: Sound, Affect, and the Ecology of Fear* (Cambridge, MA: MIT Press, 2009), 15–26.

58. Mirko M. Hall and Joshua Gunn, "Killing Them Loudly: Rhetorics of Sonic Torture," presentation to the Rhetoric Society of America, Minneapolis, May 2010.

59. Peter Worthington, "Look Ahead to Bush's 'New World Order,'" *Financial Times*, January 23, 1991.

60. Daryl G. Press, "The Myth of Air Power in the Persian Gulf War," *International Security* 26, no. 2 (2001): 6.

61. Friedrich A. Kittler, *Gramophone, Film, Typewriter*, trans. Geoffrey Winthrop-Young and Michael Wentz (1986; Stanford, CA: Stanford University Press, 1999), 100–102.

62. Rick Anderson, *Crusade: The Untold Story of the Persian Gulf War* (Boston: Houghton Mifflin, 1993), 186.

63. Pew Research Center, "Modest Bush Approval Rating Boost at War's End," April 18, 2003, Survey Reports, *Pew Research Center*, http://people-press.org/reports/display.php3?ReportID=182.

64. Scott A. Cooper, "Air Power and the Coercive Use of Force," *Washington Quarterly* 24, no. 4 (2001): 89.

65. Gordon R. Mitchell, "Placebo Defense: Operation Desert Mirage? The Rhetoric of Patriot Missile Accuracy in the 1991 Persian Gulf War," *Quarterly Journal of Speech* 86, no. 2 (May 2000): 122.

66. Shehla Burney, "Horror of War Quickly Fades from Western View," *Toronto Star*, May 17, 1991.

67. Ibid.

68. Tom Kelly, "Lessons Can Be Learned from the Gulf War," *Calgary Herald*, April 23, 1992.

69. Bosah Ebo, "War as Popular Culture: The Gulf Conflict and the Technology of Illusionary Entertainment," *Journal of American Culture* 18, no. 3 (Fall 1995): 19.

70. Jamieson, "Prof. Kathleen Hall Jamieson on Tony Schwartz."

71. Chris Hayward and Charles Rondeau, "Witness for the Execution," *Get Smart*, CBS, February 6, 1970.

72. Republican National Committee, "The Stakes," 2006, FactCheck.org, *Annenberg Public Policy Center of the University of Pennsylvania*, http://www.factcheck.org/video/RNCStakeshi.wmv, WMV.

73. Jay Heinrichs, "How to Talk like Bush," 2005, *Figures of Speech, Figaro*, http://www.figarospeech.com/talk-like-bush.

74. See Michael Kranish, "GOP Ad Puts Focus on Terror, E-mailed Pitch Aims to Energize Party Activists," *Boston Globe*, October 20, 2006.

75. Susan Page and Jill Lawrence, "War, Economy Sway Race; Democrats Hope

to Ride Wave of Voter Discontent into Control of Congress," *USA Today,* October 24, 2006.

76. See James P. Pfiffner, "Did President Bush Mislead the Country in His Arguments for War with Iraq?" *Presidential Studies Quarterly* 34, no. 1 (March 2004): 25–46.

77. "GOP Terrorism Ad Sparks Democratic Furor," *Cable News Network,* October 20, 2006, www.cnn.com/2006/POLITICS/10/20/gop.ad/index.html; and "'Daisy' Redux: RNC Ad Suggests Voting for Democrats Carries Risk of Nuclear Incineration," October 20, 2006, FactCheck.org, *Annenberg Public Policy Center of the University of Pennsylvania,* http://www.factcheck.org/article457.html 2.

Chapter 6. On Sound Criticism

1. Patrick Feaster and David Giovannoni, "The Swearing Machine," in *Actionable Offenses: Indecent Phonograph Recordings from the 1890s* (booklet included with compact disc) (Champaign, IL: Archeophone Records, 2007), 5.

2. Andrew C. Hansen, "Reading Sonic Culture in Emerson's 'Self-Reliance,'" *Rhetoric and Public Affairs* 11, no. 3 (Fall 2008): 419–20.

3. Charles Batchelor, "Charles Batchelor's Phonautograms Metropolitan Elevated Railroad from 40 Feet Away (1878 Phonautogram)," 1878, *First Sounds,* http://www.firstsounds.org/sounds/batchelor.php (accessed July 21, 2010).

4. Tim C. Fabrizio, George F. Paul, and Aaron Cramer, "A Dialogue on 'The Oldest Playable' Recording," *ARSC* 33, no. 1 (Spring 2002): 77–84; and Patrick Feaster and Stephan Puille, "Letters to the Editor: Dialogue on the Oldest Playable Recording," *ARSC* 33, no. 2 (Fall 2002): 237–42.

5. Aaron Cramer with Allen Koenigsberg, "The World's Oldest Recording: Frank Lambert's Amazing Time Machine," parts 1 and 2, *Antique Phonograph Monthly* 10, nos. 2 and 3 (1992), http://www.collectorcafe.com/article_archive.asp?article=670&id=1504 and http://www.collectorcafe.com/article_archive.asp?article=670&id=1503.

6. Richard Bauman and Patrick Feaster, "'Fellow Townsmen and My Noble Constituents!': Representations of Oratory on Early Commercial Recordings," *Oral Tradition* 20, no. 1 (2005): 37; Susan J. Douglas, *Listening In: Radio and the American Imagination* (New York: Times Books, 1999), 41–42; and John Durham Peters, *Speaking into Air: A History of the Idea of Communication* (Chicago: University of Chicago Press, 1999), 211.

7. Doris Day, "Tic, Tic, Tic," *It's Magic, Doris Day: Her Early Years at Warner Bros.* (Rhino, 2005), track 9.

8. Theodor W. Adorno, with George Simpson, "On Popular Music" (1941), in *Essays on Music,* ed. Richard Leppert (Berkeley: University of California Press, 2002), 442–43.

9. Lewis Mumford, *Technics and Civilization* (1934; New York: Harcourt, Brace, and World, 1963), 203. See also Jacques Attali, *Noise: The Political Economy of Music* (1977; Minneapolis: University of Minnesota Press, 1992), 4.

10. Richard Leppert, "Desire, Power, and the Sonoric Landscape: Early Modernism and the Politics of Musical Privacy," in *The Place of Music*, ed. Andrew Leyshan, David Matless, and George Revill (New York: Guilford, 1998), 292.

11. Gilles Deleuze and Felix Guattari, *A Thousand Plateaus: Capitalism and Schizophrenia,* trans. Brian Massumi (1980; Minneapolis: University of Minnesota Press, 1987), 313.

12. Allan Bloom, *The Closing of the American Mind: How Higher Education Has Failed Democracy and Impoverished the Souls of Today's Students* (New York: Simon and Shuster, 1987), 71.

13. Bertolt Brecht, "Radio as an Apparatus of Communication," *Brecht on Theatre: The Development of an Aesthetic,* trans. John Willett (New York: Hill and Wang, 1964), 52.

14. Balilla Pratella, "Manifesto of Futurist Musicians" in *Futurist Manifestos,* ed. Umbro Apollonio, trans. Robert Brain et. al. (1910; Boston: MFA, 2001), 37.

15. Rick Altman, "The Material Heterogeneity of Recorded Sound," *Sound Theory Sound Practice,* ed. Altman (New York: Routledge, 1992), 16.

16. R. Murray Schafer, *The Soundscape: Our Sonic Environment and the Tuning of the World* (1977; Rochester, VT: Destiny, 1994), 55–56; Alain Corbin, *Village Bells: The Culture of the Senses in the Nineteenth-century French Countryside* (New York: Columbia University Press, 1998); and Mark M. Smith, *Listening to Nineteenth-century America* (Chapel Hill: University of North Carolina Press, 2001), 267.

17. Bruce R. Smith, *The Acoustic World of Early Modern England: Attending to the O-Factor* (Chicago: University of Chicago Press, 1999), 56.

18. Martin Shingler and Cindy Wieringa, *On Air: Methods and Meanings of Radio* (London: Arnold, 1998), 64.

19. Kathleen O'Toole, "Computers with Voices: Students Explore How People Respond," *Stanford University News Service,* July 25, 2000, 30.

20. Quoted in Alex Ross, *The Rest Is Noise: Listening to the Twentieth Century* (New York: Farrar, Straus, and Giroux), 102.

21. James Baldwin, "Sunny's Blues," in *The Oxford Book of American Short Stories,* ed. Joyce Carol Oates (1957; New York: Oxford University Press, 1992), 439.

22. Robert L. Mott, *Radio Sound Effects* (Jefferson, NC: McFarland, 1993), vii.

23. Richard J. Hand, *Terror on the Air: Horror Radio in America, 1931–1952* (Jefferson, NC: McFarland, 2006), 32.

24. Ibid., 26.

25. See Walter J. Ong, *The Presence of the Word: Some Prolegomena for Cultural and Religious History* (New Haven: Yale University Press, 1967), 124.

26. Henry David Thoreau, *Walden and Other Writings of Henry David Thoreau,* ed. Brooks Atkinson (1937; New York: Modern Library, 1965), 105.

27. Luigi Russolo, *The Art of Noises,* trans. Barclay Brown (New York: Pendragon, 1986), 23.

28. Theodor W. Adorno, *The Philosophy of Modern Music*, trans. Anne G. Mitchell and Wesley V. Blomster (1949; New York: Seabury Press, 1973), 198.

29. Attali, *Noise*, 35.

30. Jonathan Sterne, "Sounds like the Mall of America: Programmed Music and the Architectonics of Commercial Space," *Ethnomusicology* 41, no. 1 (Winter 1997): 22–50.

31. Mumford, *Technics and Civilization*, 203.

32. Don Ihde, *Sense and Significance* (Pittsburgh, PA: Duquesne University Press, 1973), 32, 38, 47–62.

33. Ong, *Presence of the Word*, 114, 130, 164.

34. Jacques Derrida, "The Voice That Keeps Silent," in *Speech and Phenomena: And Other Essays on Husserl's Theory of Signs*, trans. David B. Allison (Evanston, IL: Northwestern University Press, 1973), 82, 87.

35. Didier Anzieu, "The Sound Envelope," in *The Skin Ego*, trans. Chris Turner (New Haven: Yale University Press, 1989), 157.

36. Steven Feld, "Places Sensed, Senses Placed: Toward a Sensuous Epistemology of Environments," in *Empire of the Senses: The Sensual Culture Reader*, ed. David Howes (New York: Berg, 2005), 179–91.

37. Stephen Connor, *Dumbstruck: A Cultural History of Ventriloquism* (New York: Oxford University Press, 2000), 34.

38. Joshua Gunn, "On Answering Machines and the Voice Abject," *Liminalities* 3, no. 1 (2007), http://liminalities.net/3-1/machine/machine.htm.

39. Mladen Dolar, *The Voice and Nothing More* (Cambridge, MA: MIT Press, 2006), 147–48.

40. Philippe-Joseph Salazar, *Le Culte de la voix au XVIIe siècle* (Paris: Honoré Champion, 1995), 14.

41. Wes Folkerth, *The Sound of Shakespeare* (Routledge: New York, 2002), 38–43; 38–39.

42. Ibid., 41, 43.

43. Joel Dinerstein, *Swinging the Machine: Modernity, Technology, and African American Culture between the World Wars* (Boston: University of Massachusetts Press, 2003), 69–75.

44. David Copenhafer, "Invisible Music (Ellison)," in *Sonic Interventions*, ed. Sylvia Mieszkowski, Joy Smith, and Marijke de Valck (New York: Rodopi, 2007), 171–92.

45. David H. Culbert, "This Is London: Edward R. Murrow, Radio News, and American Aid to Britain," *Journal of Popular Culture* 10, no. 1 (Summer 1976): 28–37.

46. Roman Jakobson, *Six Lectures on Sound and Meaning* (Cambridge, MA: MIT Press, 1978); and Roland Barthes, *Image, Music, Text*, trans. Stephen Heath (New York: Hill and Wang, 1977), 149–54, 179–89.

47. Rudolf Arnheim, *Radio*, trans. Margaret Ludwig and Herbert Read (1936; London: Faber and Faber, n.d.).

48. A. H. Cantril and G. W. Allport, *The Psychology of Radio* (New York: Harper and Brothers, 1935).

49. Theodor W. Adorno, "Radio Physiognomics," in *Current of Music: Elements of a Radio Theory*, ed. Robert Hullot-Kentor (2006; Malden, MA: Polity, 2009), 72.

50. See, for example, Rush Limbaugh, "Obama Says He's Not a Bolshevik, Communist Party Thinks Otherwise," *Rush Limbaugh Show*, January 29, 2010, http://www.rushlimbaugh.com/home/daily/site_012910/content/01125111.guest.html.guest.html, accessed August 4, 2010; Glenn Beck, "Obama Birth Certificate 'Horrible Forgery,'" *Glenn Beck Program*, July 22, 2008, http://www.glennbeck.com/content/articles/article/198/12701/; and Thomas Frank, *What's the Matter with Kansas* (New York: Holt, 2005), 263.

51. "Turn Glenn Beck Off—Bob Inglis (R-SC)," Townhall Meeting, Boiling Springs, SC, Youtube, August 6, 2009, http://www.youtube.com/watch?v=yMs47TSA0M4, accessed August 9, 2010. See the response at Glenn Beck and Bill O'Reilly, "O'Reilly Factor: Beck Unplugged," *Fox Network*, August 10, 2010, http://video.foxnews.com/v/3937792/beck-unplugged, posted December, 2009, accessed August 4, 2010.

52. When I was a congressional aide to Representative Leslie Byrne (D-VA) from 1993 to 1994, I was amazed by Representative Castle's calm demeanor and quick mind and can't remember him stumbling over a word.

53. "Mike Castle on Barack H. Obama Birth Certificate," *Youtube*, July 10, 2009, http://www.youtube.com/watch?v=9V1nmn2zRMc. See Brian Stelter, "A Dispute over Obama's Birth Lives On in the Media," *New York Times*, July 25, 2010.

54. Ibid.

55. John Durham Peters, "The Uncanniness of Mass Communication in Interwar Social Thought," *Journal of Communication* 46, no. 3 (1996): 119.

56. Theodor W. Adorno, "Analytical Study of the NBC 'Music Appreciation Hour,'" *Musical Quarterly* 78, no. 2 (Summer 1994): 353.

57. Walter Benjamin, "The Work of Art in the Age of Mechanical Reproduction," in *Illuminations: Essays and Reflections*, ed. Hannah Arendt, trans. Harry Zohn (New York: Schocken, 1969), 217–52.

58. Ibid., 70.

59. Ong, *Presence of the Word*, 116.

60. Walter Benjamin, "The Task of the Translator," trans. Harry Zohn, in *The Translation Studies Reader*, ed. Lawrence Venuti (London: Routledge, 2000), 19.

61. Barthes, *Image, Music, Text*, 179, 153.

62. Benjamin, "The Task of the Translator," 20.

63. Peg Rawls, "Sonic Envelopes," *Senses & Society* 3, no. 1 (2008): 61–76.

64. Altman, "Material Heterogeneity of Recorded Sound," 18.

65. Shai Burstyn, "In Quest of the Period Ear," *Early Music* 25, no. 4 (November 1997): 693; Michele Hilmes, "Is There a Field Called Sound Culture Studies? And Does It Matter?" *American Quarterly* 57, no. 1 (March 2005): 249; and Kaja Silverman,

The Acoustic Mirror: The Female Voice in Psychoanalysis and Cinema (Indianapolis: Indiana University Press, 1988), 31.

66. Benjamin, "Task of the Translator," 16.

67. Peter Szendy, *Listen: A History of Our Ears,* trans. Charlotte Mandell (2001; New York: Fordham University Press, 2008), 50.

Index

accent, 13, 17–21
acousmatic voice, 121, 130, 148
acousmêtre, 121
Adorno, Theodor W.: and Beethoven, Ludwig von, 6, 8; on dangers of sound, 147–48, 150–51; and Hitler, Adolf, 62; on listening, 7–9, 84–85, 138–39; on radio, 74, 104; on sonic studies, 10, 144
Adventures of Huckleberry Finn, 31–32
air-raid siren, 106–16, 140
Allport, G. W., 8, 61–62, 97–98, 147
alogon, 139
Al Qaida, 129
Altman, Rick, 152
ambulance siren, 140
American Dialect Society, 95
American Federation of Labor, 23
Amos 'n Andy (radio program), 14, 80, 102–4
amplifier, 124
Anderson, Benedict, 50
Andrews, Sybil, 64–66, 69
Antony and Cleopatra (play), 146–47
Anzieu, Didier, 67, 145
applause. *See* clapping
Ardoin, Amédé, 82
Aristotle, 127
Armstrong, Louis: and Oliver, King, 81, 86–87; and Rodgers, Jimmie, 76, 82; and *What Did I Do To Be so Black and Blue*, 85, 99

Arnheim, Rudolph, 84–85, 104, 138, 147
atomic bomb, 107, 111–12, 137. *See also* nuclear bomb
Attali, Jacques, 144

Bacon, Francis, 3
Baldwin, James, 141–42
Balkan War, 60–61
Ballot, Christian Hendrik Diederik Buys, 118
Barthes, Roland, 147, 151
Batchelor, Charles, 132
Bauhaus school, 85
Baum, L. Frank, 31, 83
Baxandale, Michael, 8
Beck, Glenn, 149
Beethoven, Ludwig von, 6, 8–9, 11, 144, 151
Benjamin, Walter, 50–51, 151–52
Berger, John, 8, 11
Bergson, Henri, 57, 152
Berkeley, Busby, 91
Berton, Vic, 87
Bin Laden, Osama, 129, 131
Black, Charles L., 86–87
blackface minstrelsy, 102
Blanchard, Frederic Mason, 33
blindness, 80, 85
Blind Willie Dunn and his Ginbottle Four, 78, 85–86
Bloom, Alan, 139
blues: and air-raid sirens, 106–7; and blind-

184 · INDEX

ness, 85–86; and early recordings, 81; and industrial noise, 49, 51, 54; and Jimmie Rodgers, 76; and memory, 141–42; and race, 76, 77, 79–82, 86, 94, 97; and sonorous envelope, 70–73; and synesthesia, 9–10. See also Johnson, Lonnie; Lang, Eddie; Rodgers, Jimmie; White, Booker
Blues Brothers (film), 115
Boccioni, Umberto, 14, 58, 64–65
Boston accent, 120
Bourdieu, Pierre, 24
Boyd, Phoebe, 14, 95–97, 101, 137–38
Bradbury, Ray, 117
Bradley, Jess, 66
Braque, Georges, 57, 64
Brecht, Bertolt, 139
Brooklyn Dodgers, 77
Brown, Treg, 116
Browning, Robert, 83
Brunswick Records, 79
Bryan, William Jennings, 23, 25, 30
Bugs Bunny, 122
Burch, John, 122
Burke, Kenneth, 6, 61
Burney, Shehla, 126
Bush, George H. W., 24, 125, 127
Bush, George W., 128–29

cadence, xi, 2–3, 60, 155n4
Café Society, 94, 99
Cage, John, 65
Calvert, Louis, 32
Cantril, A. H., 8, 61–62, 97–98, 147
Capenhafer, David, 147
Carmichael, Hoagy, 86
Carra, Carlo, 58–59, 64
cartoons: and Cleveland, Grover, 30; and McKinley, William, 30, 34; political, 26, 35, 42; and Roosevelt, Theodore, 37–40; and Taft, William Howard, 42–44; and Warner Brothers, 11, 13, 113–14, 116–19. See also Duck & Cover, Fast and Furryous, Road Runner, Wile E. Coyote
Cash, Johnny, 76
Castle, Mike, 149–50
Central Intelligence Agency, 123–24
Chaplin, Charlie, 61, 63–64, 70, 99
Cheney, Dick, 128
Chicago, 21, 51, 53, 71, 87, 88, 89
Chion, Michel, 113, 120–21, 148
Chocolate Dandies, 83

church bells: and memory, 142; and propaganda, 62–63; and song, 96; and soundscape, 65–66, 71, 140; and time, 49, 53, 54
Cicero, 97
clapping, 61–62
Clark, Solomon Henry, 33
classical music, 5, 8, 16, 69
Cleveland, Grover, 26, 28–31
Clinton, William Jefferson, 24
clock: as disturbance, 14, 47–55, 73, 128–30; and talking, 133–37
Cockaday, Laurence Marsham, 74
cognitive dissonance, 79, 88, 91
cold war, 108–18
Colores, 97–100
Columbia Broadcasting System, 2, 10, 108
Columbia Phonograph Company, 23, 80, 81
comedy, 16, 17, 20, 22, 61–62
Committee for Public Information (U.S.), 110
Comstock Laws for the Suppression of Vice, 16, 18
Condron, Eddie, 87
Connor, Steven, 62, 145
Corbin, Alain, 140
Correll, Charles, 102–4
Creath, Charles, 82, 89
Creole Jazz Band, 81
cubism, 47, 57
Culbert, David. H., 147
cultural capital, 94

Daisy (advertisement), 15, 119–23, 128, 130–31
Davis, Rebecca Harding, 55–56, 68
Day, Doris, 70, 137
Delaware, 149
Deleuze, Gilles, 55, 68, 139
DeLuca, Kevin Michael, 11, 138
Democrats, 128, 130–31
Derrida, Jacques, 145
Detroit, 71, 73
dialect, 13, 17–21, 46
Dickens, Charles, 55–56, 68
Dinerstein, Joel, 55, 72, 147
Disney, Walt, 117, 122
displacement, 101
documentary film, 114
Dolar, Mladen, 101, 145
Doppler, Christian, 118
Doppler effect, 71, 118

Douglas, Susan J., 4, 9, 103
Doyle, Arthur Conan, 83
Doyle Dane Bernbach, 119
Dracula, 83–84
Duck and Cover (documentary), 15, 112–13, 119, 125
Dunlap, Orrin E., 2

Edison, Thomas A., 16, 50, 132, 133
Eisenhower, Dwight D., 114, 124
Eisler, Hanns, 65
Elkins, James, 11
Ellington, Duke, 90, 99
Ellison, Ralph, 48, 85, 99, 147
Emerson, Ralph Waldo, 12
enargia, 80
Engel, Carl, 82
England, 108–10, 146
enthymeme, 87, 114, 122–23, 125, 127
Erhart, Samuel D., 42
Erskine, John, 1, 84–85

Fast and Furry-ous (cartoon), 116
Faulkner, Anne Shaw, 94
Federal Civil Defense Administration, 111–12, 124
Feld, Steven, 6, 145
Ffrangcon-Davies, David, 37
film, 113–14, 116; as object of scholarly interest, 5–6; and propaganda, 48, 62, 108, 111, 117, 119, 127. See also *Blues Brothers*, *Duck and Cover*, *Forest Gump*, *Metropolis*, *Triumph of the Will*
Finnegan, Cara, 21–22
First Sounds, 133, 136
Fitzgerald, F. Scott, 86, 105
Fletcher Henderson's Orchestra, 82, 89–90
Folkerth, Wes, 146–47
Ford, Henry, 50
Forrest, Helen, 93
Forrest Gump (film), 115
France, 12
Freeman, Buzz, 87
Freud, Sigmund, 100
futurism, 47, 58–61, 63–64, 143, 148

gaze, the, 9
Geiger counter, 137
Gennett Phonograph Company, 80–81, 83
Germany, 7, 48, 61–63, 108, 123, 148
Gibson, Clifford, 81

Gillam, Bernard, 30
Goebbels, Joseph, 62, 148
Goldwater, Barry, 120, 122–23
Gompers, Samuel, 23
Goodman, Benny, 82, 88–89, 91, 92–93, 99
Gosden, Freeman, 102–4
gravitas, 26, 30, 35
Great Britain. *See* England
Griffith, D. W., 45, 103
Grossberg, Lawrence, 6, 51
Guattari, Felix, 55, 68, 139
guitar, 48, 71, 73; in art, 57–59, 65
Gunn, Joshua, 8, 145

Hall, G. Stanley, 33
Hall, Stuart, 6–7
Hampton, Lionel, 91
Handy, W. C., 70
Hanna, Marcus A., 34–35
Hardy, Thomas, 55–56, 68
Hariman, Robert, 11, 114–15
Harrison, Benjamin, 30, 34
Hawkins, Coleman, 89
heartbeat, 67, 130, 137, 144
Heinrichs, Jay, 129
Hilmes, Michelle, 4, 7
Hitchcock, Alfred, 141
Hitler, Adolf: and sound technology, 14, 62–63, 148; and voice, 2, 48, 61–62
Hobart, Garrett, 34
Holiday, Billie: and Sinatra, Frank, 137; and *Strange Fruit*, 14, 80, 98–101; and voice, 5, 92–94
Hollywood, 76, 91, 103, 117
Hollywood Hotel, 91
Hollywood Independent Citizens Committee of the Arts, Sciences, and Professions, 117
homogenous empty time, 50
Hooke, Robert, 3
Hoover, Herbert, 10
hot air balloons, 30–31
Hughes, Langston, 87–88, 106–8
Hunting, Russell, 16–21, 33, 70
Huston, Walter, 103

identification, 107–9
Ihde, Don, 145
impressionism, 141
Inglis, Bob, 149
instructional style. *See* plain speaking

integration, 105; sonic, 15, 77–79, 81, 85–91; visual 76–77, 79, 86, 91, 94; vocalic, 92–94
Invisible Man, 85, 147
Invisible Scarlet O'Neil, 80, 104–5
Irving, Henry, 29, 32, 35

Jakobson, Roman, 147
James, William, 83
Jamieson, Kathleen Hall, 127
jazz: and early recordings, 80–81; and modernity, 70; in New York City, 85, 99; and race, 76–77, 78–79, 81–83, 86–94, 105; and sonorous envelope, 72–73, 147; and time, 50. *See also* Armstrong, Louis; Carmichael, Hoagy; Forrest, Helen; Goodman, Benny; Holiday, Billie; Morton, Jelly Roll; Oliver, King; Shaw, Artie; Tilton, Martha; Wilson, Teddy
Jean Goldkette's Orchestra, 82
Jim Crow, 79, 81, 103
Johnson, Blind Willie, 85
Johnson, Lonnie, 86
Johnson, Lyndon Baines, 15, 119–23, 124, 130
Jones, Chuck, 15, 114, 116–19
Jordan, Winthrop D., 77

Kant, Immanuel, 51, 108, 123, 139
Kelly, Tom, 126–27
Kennedy, John F., 121
Kenney, William Howland, 4
Keppler, Joseph, Jr., 30, 34
Keppler, Joseph, Sr., 26, 34
Kippling, Rudyard, 83
Kirby, Edward Napoleon, 39
Kruppa, Gene, 91
KUBARK, 123–24
Ku Klux Klan, 78, 88

Ladd's Black Aces, 83
La Guardia, Fiorello, 111
Lambert, Frank, 133–36
Lang, Eddie, 86
Lang, Fritz, 48
laugh track, 17, 142
Law, Frederick Houk, 45
Lawrence, D. H., 68–69, 75
Lee, Sonny, 82, 89
Levine, Cornelia, xi
Levine, Lawrence W.: on popular culture, 21, 27, 77; and scholarship, 5; and teaching, ix–xi, 12–13, 15, 131

Library of Congress, 95
Life, 93–94
Lights Out (radio program), 2
Limbaugh, Rush, 150
Lincoln, Abraham, 103: and appearance, 21–22; and voice, 45–46, 91
locomotive: and comfort, 54, 68–69; as disturbance, 14, 47–50, 53–55, 58, 132, 143; and Doppler effect, 118; incorporated into music, 48, 70–73, 147; and modernity, 48, 50, 70; and time, 49–50
Lomax, John, 95
London, 15, 108–10
London After Dark (radio program), 109–10
Los Angeles, 87, 103
Lothar, Rudolph, 101
loudspeaker, 14, 61, 148
Louis, Joe, 48
Lucaites, John Louis, 11, 114–15

Mann, Thomas, 68–69, 74
Marable, Fate, 81
Margulis, Charlie, 89
Marinetti, Filippo Tommaso, 58–61, 67
Marx, Karl, 50, 144
McCarthy, Joseph, 114, 117, 119
McGee, Dennis, 82
McKinley, Ida, ix
McKinley, William, 20–21, 29–30, 34
McLuhan, Marshall, 4, 113–14, 121
McTell, Blind Willie, 85
Meader, Emma Grant, 94
Meeropol, Abel, 99
Melville, Herman, 55–56, 68
Memphis, 71, 73
Memphis Five, 83
Mercury Theatre (radio program), 2
Metropolis (film), 48, 56, 70, 72
Mezzrow, Mezz, 87, 94
microphone, 14, 48, 63, 81, 92, 127; history of, 78, 81
Midwestern United States, 88–89
Milhaud, Darius, 141
military-industrial complex, 124–25
minstrel shows, 16, 27, 101, 103
Mississippi, 47, 58, 71–73
Moholy-Nagy, Laszlo, 85
Monday Night at the Movies, 119–20
Morton, Jelly Roll, 78, 81
Mumford, Lewis, 49, 70, 144
Murrow, Edward R., 15, 108–10, 119, 147

music: and consumption, 6, 144; and deep reading, 8–9; and Doppler effect, 118; and invisibility, 104; and noise, 57–58; and persuasive effects, 141–42; and race, 14, 76–94, 97–99, 105; and rhetoric, 38; and sonorous envelope, 14, 48, 62–63, 65–75, 147; under-studied, ix, 4–5, 139, 151; and unreason, 3, 6, 51, 123, 138–39, 153; and voice, 37. *See* blues; classical; guitar; jazz; old-timey; piano; shawm; zydeco
musical notation, 139
Mussorgsky, Modest, 141

National Broadcasting Corporation, 119–20, 131
National Opinion Research Center, 111
National Phonograph Company, 23, 80
National Security Act, 111
Native Son, 14, 47, 51–55
Nazis, 62–63, 123, 141
New Orleans Rhythm Kings, 78, 81
New York City, 34, 85, 87, 88, 90, 99, 108
New York City Elevated Railroad, 132
Nixon, Richard, 117
noise: and bombs, 107, 116, 118, 120; consumption of, 104; as disturbance, 3, 47–49, 51–60, 68, 73, 132, 143, 148; and modernity, 14, 47, 63; and persuasion, 3–4, 108, 124; and sonorous envelope, 60, 62–75, 140–41, 150; and torture, 123, 141; as unread, 4, 143, 144
Noriega, Manuel, 124
nuclear bomb, 116, 119, 128, 130. *See also* atomic bomb

Oakley, Helen, 89
Obama, Barack, 149–50
Oboler, Arch, 2
Odysseus, 106
Office of Civilian Defense, 110
Office of War Information, 111
official language, 20, 24, 31–32, 94, 104–5
Ogren, Kathy A., 78
Okeh Records, 80–81
Old Gold Radio Program, 93
old timey music, 80, 81
Oliver, King, 81, 86–87
Ong, Walter, 92, 95, 101, 121, 145, 151
onomatopoeia, 36, 58, 60, 70, 99
Opper, Frederick Burr, 34
organ, 142

orotund style, 27–34, 39, 45
Ott, Edward Amherst, 33

Patchwork Girl of Oz, 83–84
Pentagon, 125–27
Pepsodent, 103–4
Peretti, Burton W., 78
period ear, 3, 10
Persian Gulf War, 124–27
Pertwee, Ernest, 37
Peters, John Durham, 150
Peterson, Charles, 90, 93
Pettenger, William, 45
phenomenology, 144–46
phonautogram, 132
phonautograph, 12
phonograph: automatic, 16–21, 28; and Hunting, Russell 16–21; and integration, 79–80, 87, 92, 94; introduction of, 32, 83, 132, 136–37; invisibility of, 83–85, 101, 105; and neglected history of, 4; and political campaign recordings, 23, 28–30, 44; popularization of, 20, 23, 98; portable, 95; and studio audience, 17, 61; and time, 50
physiognomy, 9, 148
piano, 37–38, 78, 100
Picasso, Pablo, 14, 57–58, 59
Picker, John M., 4
plain speaking, 27, 31–41
Plato, 3–5, 51, 108, 123, 139, 153
Pledge of Allegiance, 150
Plunkitt, George Washington, 24–25
popular music, 5, 6, 16, 38, 78, 79, 138, 144. *See also* blues; jazz; old timey music; zydeco
pornography, 13, 16, 20
Power, Cyril, 64–65, 74
Pratella, Balilla, 139
Pray, Sophie A., 94
presidency, 13, 21–22, 26
Princeton Radio Project, 7, 147
Project East River, 111–12
propaganda, 111, 119, 127
pseudochromaesthesia, 10
psychoanalysis, 3, 80, 144–46
psychological operations, 125
Psychology of Radio, 8, 61–62, 97–98
Pygmalion, 35

race records, 80
radio: and blindness of, 86; dangers of, 109,

123, 138; identification through, 110, 115; invisibility of, 79–80, 83–84, 101–5, 121; and neglected history of, 4–5, 76–77; persuasive power of, 1–3, 10, 46, 109–10, 123, 147–51; popularization of, 98; and propaganda, 61–62; and race, 83, 85–88, 92–94; and sonorous envelope, 67, 69, 73–75; and studio audience, 17. See also Alport, G. W.; *Amos 'n Andy*; Cantril, A. H.; Hitler, Adolf; Murrow, Edward R.; Roosevelt; Franklin; *Shadow, The*; *War of the Worlds*; *Weird Circle*
radioactivity, 137
Rawls, Peg, 152
Reagan, Ronald, 127
Republican National Committee, 122, 128–29
Riefenstahl, Leni, 62–63
Road Runner (cartoon), 11, 15, 116–19, 126
Robinson, Jackie, 76–77, 91
Rodgers, Jimmie, 76, 82, 86
Rogers, William A., 40, 42
Roosevelt, Eleanor, 111
Roosevelt, Franklin Delano: and cadence, xi, 3; first inaugural address, 1–2; voice, 10–11, 46
Roosevelt, Theodore, 20–21; in cartoon, 34, 42; and *The Right of the People to Rule*, 14, 35–38; and voice, 24–25, 35–41, 44
Russolo, Luigi, 58, 65, 143

Salazar, Philippe-Joseph, 146
Sandlands, John Poole, 27
Scarry, Elaine, 68
Schafer, R. Murray, 7, 70, 113, 140
Schmauk, Theodore Emanuel, 37
Schoenberg, Arnold, 14
School of the Americas, 124
Schrimpf, George, 64
Schwartz, Tony, 119–23, 124, 128
Scott de Martinville, Édouard-Léon, 12
Seneca the Elder, 97
Shadow, The (radio program), 80, 102, 104
Shakespeare, William; dialect, x, 21, 28, 29, 32, 36, 135; and popular culture, 27, 32, 34–35; and shawm, 7, 146
Shaw, Artie, 87, 88, 92–93
Shaw, George Bernard, 35
shawm, 7, 146
Sherlock Holmes, 51, 84
Shurter, Edwin Du Bois, 33

Sinatra, Frank, 137
Skinner, Otis, 32
Smith, Brainard Gardner, 33, 34
Smith, Mamie, 80
Smith, Mark M., 7, 15, 140
sonic vernacular, 105
Sonny's Blues (short story), 141–42
sonorous envelope, 61, 96, 121, 145, 152; danger of, 149–50; explained, 67–68; and heartbeat, 130; and music, 67–75
sound effects, 3, 142
soundscape, 49, 68, 71, 75, 107, 145
sound wave, 74–75
South Carolina, 149
Southern United States, 21, 71, 88–91, 95, 106
Soviet Union, 123, 137, 141
Spain, 106
Special Streamline (song), 48, 71–73
Spencer, Len, 45
Stakes, The (advertisement), 128–31
Starr Piano, 80
Sterne, Jonathan, 7, 144
Stoker, Bram, 83
Strange Fruit (song), 14, 80, 98–101
Stravinsky, Igor, 14, 47, 65–66, 84–85
studio audience, 17, 61–62
syncopation, 92, 94, 100
synesthesia, 9–10, 58, 97, 107
Szendy, Peter, 153

Taft, William Howard, 24–25, 39, 42–44
Taylorism, 50
Tea Party (political movement), 149–51
telephone: invisibility of, 101; popularization of, 98
television, 6; dangers of, 123; dominates scholarship, 110; invisibility of, 121; and laugh track, 17, 142; and racial integration, 76–77. See also *Daisy*; *Stakes, The*
Tennyson, Alfred Lord, 4
Thoreau, Henry David, 97
Tic, Tic, Tic (song), 137
Tilton, Martha, 92–93
Tinfoil.com, 5, 133
torture, 123–24, 141
train. See locomotive
translation, 151–53
Triumph of the Will, 62–63
Truman, Harry S., 111–12, 114
trumpet, 10, 76, 100, 118
Twain, Mark, 31

uncanny, 80, 100–102, 105

Van Vechten, Carl, 87–88
vaudeville, 29
ventriloquism, 72, 145
Victor Talking Machine Company, 23, 76, 80, 81, 82
Victor Talking Machine Co. v. Starr Piano, 80–81
video game, 124, 126
visual metaphor, ix–x, 3, 11
visual studies, 7–8
voice: and Bryan, William Jennings, 23; and *Daisy* 120–22; and Harrison, Benjamin, 34; and Hitler, Adolph, 48, 61–63; and Holiday, Billie, 5; and Hoover, Herbert, 10; invisibility of, 80, 105; and Lambert, Frank, 134; and Lincoln, Abraham, 19, 45–46; and machines, 55, 68; and Marinetti, Filippo Tommaso, 60; and Murrow, Edward, R., 110, 119, 147; and persuasion, 4, 101–2, 110, 141, 148, 151; and presence, 80, 91–101, 137, 144–45; presidential, 14; and race, 15, 94, 102–4; and Roosevelt, Franklin Delano, 1–2, 10–11, 46; and Roosevelt, Theodore, 22, 35, 38–39; and Scott de Martinville, Édouard-Léon, 12; and self, 10–11, 79, 84, 91, 94–95; and sirens, 106; and sonorous envelope, 67–68, 121, 148–51; under-studied, ix, 3–5, 123; and unreason, 3. *See also* accent; acousmatic voice; dialect; orotund; plain speaking
Von Harbou, Thea, 48
Vorhees, Jerry, 117

Warner Brothers, 117–18
War of the Worlds (radio program), 2, 105
Washington, D.C., 40
Washington, George, 42
Weaver, Andrew Thomas, 95
Webb, Chick, 90
Weird Circle, The (radio program), 142
Welles, Orson, 2–3, 102, 104–5, 121
White, Booker (Bukka), 14, 48, 71–73
Whitman, Walt, 70, 72
Wile E. Coyote, 114, 116–19, 126
Wilson, Teddy, 81, 83, 90, 91–92
Wilson, Woodrow, 25, 39
Wolverines, the, 82
Works Progress Administration, 95
World War I, 14, 110
World War II, 15, 108–11, 149
World War III, 112, 116
Worthington, Richard, 124
Wright, Richard, 14, 47, 51–55, 71, 73, 99

zydeco, 80, 81, 82

GREG GOODALE is assistant professor of communication studies at Northeastern University and the coeditor of *Arguments About Animal Ethics*.

The University of Illinois Press
is a founding member of the
Association of American University Presses.

University of Illinois Press
1325 South Oak Street
Champaign, IL 61820-6903
www.press.uillinois.edu